Len Goodgal (arrow) of the 101st Airborne Division with the Rangers on Pointe du Hoc.

A Walk in the Sun

and other stories from the WWII Oral History web site

By AARON ELSON

Published by Chi Chi Press

Chi Chi Press
12 Washington Street
Suite 55
New Britain, CT 06051
(888) 711-8265

ISBN 978-0-9727330-5-2

About the Author

Aaron Elson began interviewing World War II veterans in 1987 at a reunion of his father's tank battalion. In 1994 he wrote "Tanks for the Memories," a collection of stories told by the veterans of the 712th Tank Battalion. The title was heisted from a bumper sticker on many of the cars in the hotel parking lot. In 1997, he launched the World War II Oral History web site. One of the first sites with significant World War II content, the stories and veterans featured on it have been used in more than a dozen documentaries and as source material in more than two dozen popular books.

Aaron's most recent book is "Semper Fi, Padre," about Mathew Caruso, a young Marine chaplain's assistant who sacrificed his life to save the chaplain during the Korean War. Mathew's brother, John Caruso, is a co-author of the book.

Books by Aaron Elson

Tanks for the Memories
Semper Fi, Padre
A Mile in Their Shoes
Nine Lives
They Were All Young Kids
The Armored Fist
The D-Day Dozen
King Kong Down
Merry Christmas in July
Semper Four
From the Cavalry to Czechoslovakia
Big Andy

www.tankbooks.com

Contents

Budd Squires with his wife and daughter.

"I Gotta Tell You Guys a story"

I was out to lunch with three veterans of A Company – my father's company – at the 1993 reunion of the 712th Tank Battalion, when one of the veterans, Budd Squires, related the following story. When the battalion newsletter arrived a few weeks after the reunion ended, it announced the passing of Squires, who went to the doctor after the reunion and was diagnosed with advanced-stage cancer. I decided then and there that rather than waste months and perhaps years looking for a literary agent or a publisher, I would self-publish "Tanks for the Memories." I have not looked back.

Budd Squires

I gotta tell you guys a story, you'll enjoy this.

We were in a little town, I don't remember where the hell it was, just a little town we fought to get in, and took the town. We were behind a building with our tanks, and the gas trucks came up and we were gassing up the tanks. But the infantry was still going through the buildings, sorting out stuff. And there was this one infantryman, he went in what was like a basement entrance, the kind you open up and go into. I was filling my tank, holding the gas can, and I watched him go in there. Pretty soon he came out, and he had his helmet in his hand and he was dragging his rifle like the goddamn war was over.

I thought he was hit, so I jumped off the tank and I said, "Are you okay? Are you okay?"

He said, "You wouldn't believe it. You'll never believe it. You would never believe it!" Just like the goddamn war was over.

There was shooting going on, and we were behind the building, so I pulled him in and said, "What the hell happened?"

He said, "You wouldn't believe it," and he just kept shaking his head.

I said, "What the hell's the matter, what happened, are you hurt?"

"No." But he went down in this building, and he came back out, and there were a couple of nuns down there, and a woman having a baby. Now that was an experience for a guy, and he was just out of it when he came out of there. He didn't give a damn if the war was over or what.

It was a hell of a time to have a baby. I remember that clearly, it's something you don't forget.

Bernie and Luenette

A Veteran of the Bulge

Bernie Barenbrugge

This is Benjamin Bernard "Boogie" Barenbrugge's account of some of his experiences as a young soldier in the U.S. Army during WWII. He was part of the 10th Armored "Tigers," who fought bravely at the Battle of the Bulge. This story was edited and submitted by Bernie's daughter-in-law, Lyn Barenbrugge. In the early days of tankbooks.com, in the late 1990s, I had a guestbook that I eventually removed because it was targeted by Russian spammers. But one of the earlier entries was from Omar Kahlgren of Crossville, Tennessee. Omar said that he was named after Omar

Billet, who was mentioned on my site. Bernie's story describes a remarkable act of compassion by Captain Billet, who later was killed in action.

PS: Why was he nicknamed "Boogie"? I don't know. It was only after this story was published on tankbooks.com that Bernie told his daughter-in-law that all the guys in his platoon had nicknames, and that his was Boogie. It might be reasonable to surmise that it was because he could dance like one of the "Boogie Woogie Bugle Boys" from Company B.

In November 1942, at the ripe old age of 21 and compliments of the draft, I was sent to the Induction Center at Fort Benjamin Harrison, Indianapolis, Indiana. I pulled kitchen KP there for a few days and was then sent to Fort Benning, Georgia, where I joined Company E of the 54th Armored Infantry Regiment. This was part of the 10th Armored Division. Here I went through about a year of intense infantry basic training. This was hell and a lot of hard work, but we also had a lot of fun drinking beer at the PX in the evenings. That helped to get me through it, and the homesickness. I really missed the old hometown of Dunkirk, Indiana, and my folks. I especially missed Luenette Turner, a girl from Muncie, Indiana, whom I had begun dating about eight months before I left.

The division then moved to Camp Gordon, Georgia. We had two more months of training and maneuvers in the mountains of Tennessee. I spent three months at a communications school at Fort Knox, where I was promoted from Buck Sergeant to Technical Sergeant and was made the Communications Sergeant for the company. This made me responsible for all communications for the Company Commander between our platoons and the battalion headquarters. This was done with either the radio or telephone equipment. Our main radio was a large unit mounted in the half-track, which we rode in during combat. It operated off the half-track battery and had a long whip-like antenna. It had a range of several miles. We also had smaller portable radios, which were carried by the platoon leaders.

In addition to the radios, we had phones in all the necessary places. These were all battery operated and connected by phone wire. They were only used when we were in a dug-in, fixed position. It was my job to string the wire between all the locations. Stringing wire at night, through a bunch of trigger-happy guys in foxholes, was a little scary at times. I had to keep all of this equipment repaired and in good working order. It was a big job, but I liked it. I rode in the company commander's half-track (two wheels in the front and tracks in the back, and it held ten guys). We had a .50-caliber machine gun mounted on a turret in the front and a .30-caliber machine gun mounted on a pedestal in the back. The guys in my company were all general, run-of-the-mill GIs. Just a bunch of average good guys, wanting to get the war over with so they could return home.

The division began to get ready to ship to Europe, so Luenette and I decided to get married before I shipped out. She traveled down to Camp Gordon by train. We were married June 26, 1944, by the post chaplain, Father John Jedlowski, in the little white frame Post Chapel No. 7. It was a short but sweet ceremony, around 6 p.m., on a Monday evening. It wasn't fancy, but it was legal. Afterwards, we all went to a hotel in Augusta for dinner. I had a three-day pass, so we also honeymooned in Augusta. Our wedding party consisted of best man John Pippert of Cincinnati, who was also my best army buddy; Augusta native Ellen Biggar was maid of honor; her aunt, Roberta Biggar, and her friend, Hazel Word, were also in the party.

John and I had met Roberta and Hazel soon after we arrived at Camp Gordon. They were two real nice elderly Southern ladies who were doing their part for the war effort by entertaining soldiers. They had a fine home and invited us in for Sunday dinners. Their favorite menu was fried chicken with all the trimmings and plenty of Mint Juleps. It sure was good! Luenette stayed and in a couple of weeks I was granted an unexpected 10-day furlough, so we went home to Indiana together. In the meantime, the division cooled off. I went back to camp and Luenette came back down to Georgia later and stayed a few more weeks until I was sent out. It was here at Camp Gordon that our regiment's name was changed to the 20th Armored Infantry, B Company.

On September 11, 1944, the 10th Armored "Tigers" (which had become our nickname) shipped out of New York. We were about 10,000 strong at that time. I was one of the poor souls on the USS General William M. Black troop ship. We slept in bunks in the lower part of the boat and it was hot and stinky down there. We ate in the ship's mess hall. The cooks, if you could call them that, were Navy men and the food was terrible. I ate very little, because I couldn't keep it down. I was seasick the whole trip. The weather was rough and the ship did a lot of bouncing around. We were all afraid because German subs were sighted near our convoy many times. Our destroyers were dropping depth charges in an attempt to destroy them. This went on for the 12 extremely long days it took us to cross the Atlantic. I don't know if any of the subs were sunk or not.

The only entertainment we had during this time were card and crap games (for those that were able). We landed at Cherbourg, France, on September 23, 1944, after one helluva rough crossing. I was never so glad to get off a boat in my whole life. We immediately pitched our tents in a big muddy field. We were here a month and spent most of our time in more training and getting our weapons, equipment and vehicles ready for combat. We did get to walk around Cherbourg some. It was a picturesque old French town. I was able to write home.

On October 24, 1944, we were on the move. We had been assigned to General George Patton's 3rd Army. We passed through Paris, where we were greeted with great joy by the French people. They gave us flowers and cider to drink. We gave them cigarettes, and candy for the kids. They were all so happy the Germans were gone. We moved on out of Paris and after that it was a continuous fight, taking one town after another.

Our first contact with the Germans was November 2, 1944, at Metz, France, and Fort Driant. Germans had occupied the fort and we exchanged artillery fire for a couple of days, until the fort was reduced to gravel and these particular Germans were no longer a problem. From there on, it was pure hell. We chased those damn Germans from town to town. They shelled us and we shelled them. We both took a lot of casualties. There was no hand to hand combat; all the killing was done by tank fire and artillery.

I guess I had a guardian angel riding on each of my shoulders. We ate K rations on the run and slept very little. There was no time to clean up. General Patton's theory was, "When you have them on the run, keep them on the run!" There was no rest for the weary. It was our job during combat to provide flank protection for the tanks from the enemy infantry. Wherever the tanks went, we went.

On December 17, we received orders to move north, as the Germans were making a large counterattack. December 18, we arrived in Bastogne, Belgium, and hit the Germans head-on. This had to be the roughest day of my service. We were completely outnumbered and surrounded. We held them off for eight miserable days. On December 26, 1944, the 4th Armored Division broke through the lines to help us. Were we glad to see them! We had run out of tank and artillery ammo. We were also out of fuel for the tanks. We were being supplied from the air by large C-47 transports parachuting in ammo, fuel and food.

I had plenty of winter clothing, so that wasn't a problem for me. Our main problem was hunger. We had no food. Between artillery barrages, we scrounged through empty houses in Bastogne. The civilians were all gone and we gathered up a lot of home-canned vegetables and meat they had left behind. We built a fire in a wood-burning stove in the kitchen of the house we were in and made a huge pot of stew. This was on Christmas Eve, and the stew was just about ready to eat and then WHAM! The Germans were bombing Bastogne. One bomb hit in our back yard. Luckily, we were in the front part. Needless to say, the stew, along with the back half of the house, was gone! We all grabbed our gear, ran across the street and went into the basement of a large Catholic cathedral. The bombing went on all night and so did our praying. The dawn finally came and the bombing stopped. The cathedral and all its occupants were spared, thank the good Lord! The Germans then dropped parachute troops on the town. There was snow on the ground and all their paratroopers were garbed in white to conceal themselves. Thanks to our crack riflemen, they were quickly eliminated. This was no doubt the worst ordeal of my entire Army career. Hunger, capture and death seemed imminent. Under these circumstances you do some serious praying and think a lot of your loved ones, while you're following orders to survive.

On January 29, 1945, we finally left Bastogne, and the Battle of the Bulge was over. After having been relieved at Bastogne, we headed back south. Our orders were to push the Germans further back into Germany. The Army engineers literally built our bridges as we moved. They got a good workout. The Germans blew every bridge across each river we had to cross. Our engineers built new bridges made of large rubber pontoons, lashed together and anchored to the river bed. They placed steel grid channels across the pontoons, spaced far enough apart to accommodate our tank tracks. I really felt sorry for these engineers, as this was a hard and dangerous job. German artillery would blow these bridges up as fast as they built them. Then our artillery had to go to work and knock out their artillery. Sometimes this went on for days. When we could, we would rest up in some little nearby town. We made three major river crossings: the Saar, Rhine and Danube. We took over 150 towns and cities. Some of the major ones were: Trier, Kaiserslautern and Mannheim.

Our company commander, Capt. Omar Billet, was killed in action April 1, 1945, at Crailsheim. Our column had come under artillery fire and the captain and his jeep driver had gone up ahead to a little white church with a tall steeple. They climbed up in the steeple, hoping to locate the enemy fire. The Germans spotted them and they laid an artillery round right on the steeple and they were both killed instantly. Their bodies were placed on doors off the church and brought back to camp on the hoods of jeeps. Capt. Billet was without a doubt the best officer we had. We all missed him very much. He was the most fearless and kindest man I ever met. I remember one night we were in foxholes on the side of a muddy hill. It was cold and pouring rain. The Germans were shelling us. The captain was walking through our area checking to see if everyone was okay. My buddy and I were in a hole together. When the Captain came by, he asked us how we were doing. We said we were doing okay, but we were cold. He took off his trench coat and threw it over us. He said, "Maybe this will help a little," and walked on.

So many of our buddies died. Our company jeep driver, Sanders, was killed; as well as, Lt. Col. J.J. Richardson, by artillery fire near Kirchen, Germany. Richardson was well liked by all the troops and was a good leader. The word was he had the biggest pair of field glasses in the division, but he never kept them in their case. He kept two fifths of Scotch in the case, to ward off the chills. Ha! We proceeded on through Germany and into Austria. The Germans had blown the road off the side of the mountain, so we had to stop. We pulled back to Murnau, Germany. It was May 1945, and thank God the War was over!

I was transferred from company headquarters to battalion headquarters and assigned to S-2 intelligence. Our job was to round up any Germans still hiding in the mountains or woods. Willie Messerschmidt, the designer of the German fighter plane, was among those we captured. Our division took over 6,000 prisoners of war. Most of these were taken toward the end of the war. They would come running out of the fields and woods to our tank column, waving white flags and yelling "Kamerad!" We would send them to the back of the column, where the foot soldiers would escort them to prisoner-of-war camps. The photos used in this document were taken on a camera that had been "confiscated" from a German POW.

We liberated a concentration camp holding several thousand Polish soldiers. We gave them all of the extra candy and cigarettes we had. Boy, were they happy! We never saw any American soldiers in camps in our area. Most of the homeowners had migrated ahead of our columns, deeper into Germany, out of harm’s way. When we reached Murnau, we received some well-deserved rest and relaxation.

We stayed in some big, fancy houses that we had taken over on the shores of Lake Staffelsee. It was a beautiful lake in the Alps, with a nice swimming beach. Almost all the civilians were in their homes in this area, as there was no fighting here. We went swimming, boating, and they even had a stable of riding horses. I was one of the lucky few from our battalion who won a two-week pass to London and Paris in a "hat drawing." We saw about all the major attractions we could. Most of the major shows and restaurants were still closed because of the war. The thing that impressed me the most in London was the areas that were hit by the German V-bombs. There was nothing left but big piles of brick as far as the eye could see. This used to be factories, stores and homes. What a disastrous sight. It is hard to imagine what these poor people went through.

For my three years' service, I did receive a few awards to help ease the pain. On a sunny day in May, our battalion assembled in a big field close to Murnau, and the commanding general of the 10th Armored Division, Major General Fay B. Pricket, presented all of us with our awards and citations. Unfortunately, there were no families present. No photographs were given to us individually. They probably all went to division headquarters. There was no reception either.

I received awards for Good Conduct, Marksmanship and Combat Infantry. I won a Bronze Star for crawling, under artillery fire, to reach Johnny Smith, a wounded member of my squad. I dragged him back to a bomb crater where several of us had taken shelter. Thankfully, his wounds weren't too serious and he was able to continue on with us. Appleton, another squad member, wasn't as lucky. We had to leave him lay for the graves people to pick up. What a shame. I was also awarded the Purple Heart for a small shrapnel wound to my right hand, received while under German artillery fire. I guess I zigged when I should have zagged. I was lucky my guardian angel was with me again! I was also awarded a

Campaign Ribbon with three Battle Stars. Our Battalion also received the Presidential Distinguished Unit Citation for our part in the Battle of the Bulge. We were all very proud of this award, and it was presented to us by Major General Anthony McAuliffe, the general who told the Germans "Nuts!" when they asked him to surrender at Bastogne.

In September 1945, I received the sad news from the Red Cross that I should come home as soon as possible, because my mother was in the hospital and was not expected to live. Her appendix had burst and gangrene had set in. It was over a month before I could get home. I wasn't able to get a plane out, because they were all being used to return doctors and medical staff and patients to the States. Since I couldn't get a plane, I had to go to France and wait for a boat. By the time I arrived home, Mother had recovered and was up doing her housework! I came home through Camp Atterbury and was processed and discharged from there in October 1945. My three years of service were over and I was home to stay!

– Edited by Lyn Barenbrugge

Just Keep Whistling

Valentine Miele

Valentine Miele, of Palisades Park, N.J., landed on Omaha Beach on D-Day with the 1st Infantry Division. This story takes place in the Huertgen Forest.

We went into Germany on the 12th of September [1944]. We went into some wooded area, and we went up this dirt road, and the Germans start moving around, so they set up the gun. My sergeant says, "Miele, get on the gun."

I got on the gun and I fired maybe ten, fifteen rounds, and they started to holler, "Cease fire! Cease fire!" And some of the Germans gave up. Then we walk up this road, and I look up and there's this big tower. It must have been in peacetime like the forest rangers have. It was about 50 feet high, made out of stone. And next to it they had a little house, two or three rooms. I guess they must have lived in there. And part of the house was knocked over. So I dug a hole – you couldn't dig much of a hole because of the roots and rocks. And we stayed there the whole night. Then in the daytime we went in the house. We went down in the basement.

It had the open doors, and the machine gun was set up there. I had a big log on the right side of the machine gun. I had the ammunition on the left. And my sergeant said, "Miele, you get on the gun." We always had one man on the gun. So it was my turn.

My sergeant says, "Miele, you better stay on the gun." So I stayed on the gun. I'm up there, and all of a sudden one of our guys comes out of the tower, and he's carrying a B.A.R. [Browning Automatic Rifle.] He comes about 25 feet away from me, and he waves to me, and then he goes down. When you go in the country they had barbed wire, they've got three strands, for the cows. But they must have had them up for the deer in the woods, because we were in a wooded area. We must have been about seven or eight miles out of Aachen, and you could hear their artillery; when they were firing you could actually hear the gun, we were so close. But you could never see it.

This guy went as far as the barbed wire, then he went over about 25, 30 yards to his right. And he went down on his hands and knees and he was crawling away, and I'm watching, and all of a sudden I hear that B.A.R. barking. And I'm watching. I'm watching. All of a sudden I see an explosion. Then I see him, he turns around, he starts to crawl away, and then another explosion. That was the end of him. And I'm just standing there watching. All of a sudden I see these guys come up the hill. They got as far as that god darned barbed wire. They must have been about 25 feet away from me. About 20 of them. That's as far as they got. Some of them got away but not all of them. Then I don't know if they threw a grenade or what. There was an explosion above my head, and my helmet came off. My head seemed to go into my shoulders. It put holes in my ammunition box and knocked the front side right off my machine gun.

Then I went down into the cellar. I said, "Wright! Wright! Put somebody else on the gun."

He said, "What's the matter?"

I said, "My neck. My neck."

He said, "You hit?"

I said, "No, I'm not hit. But my neck."

The next thing you know, a shell comes in and hits the house. Ba-Wham! Then they got plaster, dust, smoke and every damn thing. You couldn't see nothing. So I said, "Hell, I'll get out of this house." I went around, I ran right smack into the steps. I had no helmet on, put a bump on my head. I walk around the steps, and there's my sergeant. He's out.

"Miele! Miele!"

"What's the matter?"

"My legs! My legs."

I looked down at his feet. His feet were swollen. The concussion got him.

So I pick him up, throw him on my shoulder, and run up the steps. When I get up there, the house is on fire. So I go out the side door, and I run him over to the tower, put him down, and he says, "Miele, you'd better get the gun." So I had to go back in the house. I couldn't go in the way I came out because it was on fire. I had to go down in the cellar and pick up the machine gun and a couple of boxes of ammunition. I look down and there's my pack. Kaput. There was nothing to my pack. The shrapnel set it afire; everything was burnt.

We got out of there, and then I went into the tower because you couldn't stay in the cellar anymore, the house was really burning then. I put the gun down, and then the other sergeant came in from the rifle platoon. He needed somebody to carry the dead. So we went out, and we got a ladder, we had no litters. We had about a 10-foot ladder, we put two bodies on it and brought it back to the tower and laid them there. Then I went back into the tower, and the platoon sergeant asked my sergeant who his gunner was. He says, "Miele's the gunner." And he said, "Well, he don't go carrying no dead." So I stayed there.

So from there we put Wright on with the dead, and I told him, "You'd better keep whistling and singing."

He said, "Why?"

I said, "Because everybody on this jeep is dead but you."

"You dirty bastard."

Then they took us out on the 13th. They took us out in the afternoon. We lost about 33 men.

Machine Gun Hill

Darrell Petty

Darrell Petty, of Newcastle, Wyoming, was a member of the 90th Infantry Division. He was wounded three times, and served in the army of occupation as an MP. The battle he describes here took place between the Moselle and Rhine rivers in mid-March of 1945.

We ran into quite a firefight on Hill 451.

We started marching up in reserve, and we could hear digging. We knew somebody was digging in, but we didn't know who. We went up a pretty steep hill, and a guy by the name of Gene Miller and I helped another guy up, his name was Prey. He was a private first class. If I knew then what I know now I'd have known he was having a heart attack. He carried a little 536 radio, they're pretty heavy those little devils, and I took that. Miller was packing his M-1 rifle, and we were lightening the load for him as much as we could.

We got to the top of the hill, and then we sat down for a break. We could still hear these guys digging in, and it turned out it was F Company. About that time, we were just sitting there and here comes a burst of machine gun fire. And man, we whirled around and headed for cover. And this kid, this Pfc. Prey, let out a groan and collapsed. We figured he was hit. We got over the hill and then hollered "Medic!" And a guy by the name of Doc Roberts, Thomas Roberts, he was a field medic but everybody called him Doc, he and he and I crawled out there to Prey and got ahold of him and dragged him back behind cover. But he was gone. And he had a classic look, in the coloration of his face, of a person with a heart attack.

We tried to find where he was hit, but we couldn't find any blood. He didn't have a bullet mark on him. And we lost him.

Lieutenant Colonel Cleveland A. Lyttle was leading us. He was the battalion commander, and he went right up that hill with us. All we know is they said there were machine guns on the hill, well, that was standard. If we'd have known what was on it we probably wouldn't have tried to take it. They pinned F Company down and they had killed 25 men and wounded some others. There were five companies of German SS dug in on that hill, and they had 40 ground type machine guns. And F Company had just buggered into them. They weren't supposed to be there, by all the reports there was nothing there. And Lyttle came up and said, "We've got this hill to take, it's got some machine guns on it, we're gonna take it." So we called in artillery, everything they had, and boy, they were tossing them in there close to us. We had to go down the hill, under trees, and then we had to cross the valley, that's where they pinned F Company down. We went through F Company, and we headed up the hill. But when we got underneath that canopy where we could look up under there, it looked like an anthill, there were Germans running all over. Well, hey, you couldn't do anything but go forward. If we turned around and retreated, we'd have been just like F Company. So we just kept going in and they'd taught us use that march and fire, every time your right foot hit the ground, if you had a carbine you fired it, and we got good shooting from the hip. Heck, I could throw a small bucket out and fire when I throw it and hit it five or six times out of eight. That's the way we went up that hill, and that colonel, he went up the hill with us. Our watches were all synchronized when the artillery was gonna stop, so we knew what time to hit the hill. And when we got to the top, we took just one German prisoner.

He was a sergeant. He spoke English, and he said, "You damned Americans are crazy." He said, "You don't know how to fight a war. When you're fired on with full automatic weapons you're supposed to hit the ground and take cover. That other unit did and you didn't, you just kept coming at us." He said, "We couldn't get our heads up to shoot back straight."

He said we took war for sport, because we laughed at things. Sometimes it was either laugh or cry, so we laughed. And by golly, it's an awful thing to say, but we shouldn't take prisoners, because if you stand there guarding that prisoner you're gonna get shot. And some, very few of them, tried to but most of them tried to fight, but a few tried to surrender. You couldn't stand there and guard him because you're going to get shot. Plus we needed everybody up the hill. So you just had to do what you did.

We took a kid by the name of Speaks, he had a B.A.R., and I had an M-1, and old Doc Roberts was like a squad leader, he was right up there on the front end with us, and we overran the command post. A full German colonel came out and his cadre, and they were running and we opened up on them with the B.A.R. and that M-1 and it just folded them up. There was one still alive, and Roberts went down and was gonna try to help him, and I heard a Schmeisser bolt click and I hollered, "Doc! Look out!" I looked up the hill and he was taking aim on old Doc and Doc just fell down among the bodies and he sprayed him and he finished killing the German, but he didn't get Doc. Speaks and I opened up with the B.A.R. and the M-1 and we nailed that guy. And my M-1 was so hot it wouldn't quit firing, it was setting itself off.

On the way up, a kid by the name of Phyllis, in F Company, just like a girl's name, his last name was Phyllis. He was mad, and he jumped up and said, "I'm going with you!" He had gone through basic with us, and he was in F company. He shouldn't have gone with us, but he did. And halfway up the hill I got a bunch of machine gun bullets through the pant leg, and they cut him down. I thought my leg was gone, it felt like somebody knocked it off. I looked down. It was still working. But if he hadn't gone with us, a kid out of Prescott, Arizona, by the name of Billy Bacon and I, we would have run out of ammunition halfway up the hill. We went back and got his ammunition and finished it up, and when it was done, we split it and I had 18 rounds left, and we were expecting a counterattack. We were sitting there with dang little ammunition, but as it turned out, this German sergeant, he saw what was going on and he played dead. He smeared blood on his face and lay there. When we discovered he was alive, Lieutenant Kelso said, "I want to talk to him," because old Sergeant Will was about to kill him. Lieutenant Kelso said, "I want to find out what they're doing here."

They were supposed to let us bypass them, and they were going to hit us from the rear that night, and the 11th Panzer was going to hit us from the front. The 11th Panzer was a German armored unit. And of course when we found out, we called artillery in on where the 11th Panzer was gonna come in and we foiled the whole thing. We were put in for a presidential citation for that, and it was on the Army Hour and was broadcast all over the free world. And I have the little article at home where it says Colonel Lyttle and G Company of the 358 outmaneuvered and destroyed five companies of German SS, and I've had people ask me about those holes, I had seven holes in my pant leg.

My officers and everyone tried to figure out how the bullets could make those holes but miss my leg. And I got a letter from a buddy that was there, he got married after we came home and he said, "I sure would like to have you meet my wife. I've been telling her all about you and what we did over there." Then he said, "Well, not quite everything." But he said, "I even told her about the seven holes in your pant leg."

We called that Machine Gun Hill. We figured we were kind of justified in doing that.

The Falaise Gap

John Warner

John Warner was a forward observer with the 344th Field Artillery Battalion of the 90th Infantry Division. I interviewed him at the 1995 reunion of the division in San Antonio.

I've been face to face with the enemy, but it's not like the First World War; I didn't see any bayonets. You see pictures of that. I never saw a bayonet being used. Now in the Pacific it's a different story, but I'm talking about in Europe. I never saw a bayonet being used by anybody.

I remember, in a little town – in fact my friend, he went back to this town – Distroff, at the end of France, almost going into Germany. We went in, along with four tank destroyers – they went in and we went in behind them. The town had maybe 500 people. There was just one street going through it. It was late in the afternoon when we got there. We just walked right in and G Company of the 358th Infantry Regiment put their outposts out.

There were hills around the village, and during the night the Germans came in. They captured I don't know how many men out of G Company, an awful lot of men. I heard a hundred, in that neighborhood. It was about 4 in the morning, it was just gonna break dawn, and I heard all this noise.

A WALK IN THE SUN

I was in a corner building, and I was a radio operator. I stepped out on the stoop, and there's a German halftrack coming, and it's 50 or 75 yards away. They had a .20-millimeter cannon, and they were firing down the street. Hit my radio. I got an award there for calling fire in. This one TD right around the corner from us, a friend of mine from Ford City, Pennsylvania, was hit there. They knocked his tank destroyer out and he was hit with shrapnel.

I saw this halftrack, and a shell hit my radio. I hurried back in the house, and my officer, Lieutenant D'Angelo, said, "We've got to call some fire back there." So I went back out, I fixed the antenna to where it was working, and we called fire direction. And they didn't want to take my order, because they said, "You guys are all captured." Three of these TDs took off and went back. And they told them that we were surrounded, that we were all done. Twice I called, and they didn't want to take my fire order. So finally they accepted it. And we fired over there and the Germans retreated and went back out over the hill. But we were face-to-face there. I would say I was face-to-face with that halftrack.

As I was going back in this building – it was a corner building, I don't know if it had been an old hotel or what – there were two big plate picture windows in my corner, and the Germans up on the hill fired an 88. I was walking in and this captain from E Company of the 358 had his headquarters down in the cellar. So I was walking over towards it and this 88 came right through this plate glass just over my head. I was just going down the steps, I was bending, and it went over my head. You could feel the heat off it. It knocked me down in the cellar. Dust. When I came to, I thought, "Where am I?" I couldn't see a thing. It scared the hell out of me, I'll tell you. Didn't hurt me, though.

In Dillingen, Germany, we were in this house, they were in that house, it was that type of situation. We'd peek out and try to find a target, direct fire on it, and they'd do the same thing. They'd come out once in a while, back and forth between buildings, you'd see them. You were face to face in that way, but as far as me going over and getting ahold of that guy or him getting ahold of me with a bayonet, no. But it did happen, understand. There were times that did happen.

At the Falaise Gap, we had a pilot named Matthews. He was in a Piper Cub. Now this Falaise Gap was a big wide valley, and there was a hill on that side and a hill on this side, and I was on the side of one of the hills.

And right down below was all this mass of humanity and vehicles, they were trying to get out and they were trapped. Every kind of vehicle you could picture was coming up through there. And personnel. You could look over and see fifty, a hundred men. Well, he was flying around in a Piper Cub, and then he was directing fire. And one of the things that we always remember him saying is, "Stop computin' and start shootin'." We always remembered that. And he was doing a good job. But they forget, this guy on the hillside there, it was man-to-man, seeing this guy, and subject to fire. And after the day was done, he flew back to his base, got a bite to eat, a place to sleep. My day didn't get done. Come dark, you were still on the damn hill. And you ate a K-ration. Forward observers got no credit. He got a lot of credit, and I don't begrudge him, because he did a good job, understand. But his conditions when he got back were a hell of a lot different than ours. He's a nice guy, I don't begrudge him any of his glory. But the guy, the forward observer from the infantry there, he gets nothing.

What I saw in the Falaise Gap was disgusting. Revolting. I'll tell you, and I'm not ashamed to be a part of it because it was a necessity, I mean it was a job I had to do. But I don't care who you are or what type of people you are, it's hard.

After the battle was over, I went down into the Gap. Me and a friend, we walked down in that valley. And it was a mess. Everything, all kinds of vehicles, you can't believe it. I mean, of every description. People laying around, blown up. An arm over there, a leg over there, you can't believe it. Can't believe it. There was a communications halftrack, and it was blown up, just stopped there. And I was curious. I looked in the back of it to see what kind of instruments they had. They had a lot of radios, wireless and all that. And I'm standing on something ...mushy. I looked, and see it was a body. Burnt. That had been blown up and burned. It was just a mass of flesh, burned. It's revolting.

I don't know, it's hard to describe. It's just like, you could be there, that could be you. Nobody shows respect for anybody there, you're just a piece of flesh. How one man could do that to another...

Hey, when you're in war, you're in war. It's a terrible situation. But how in hell you can do something like that to another person, I don't give a damn who they are. Terrible. It's the type of thing, hey, I could tell you all day, but there's no way you could describe a picture like that.

That's the difference between a Matthews and me. He's a good man, but he didn't see that. He wasn't a party to that. He can talk to you all day, about hey, I soared around there, and I directed fire and killed fifty men. I was never proud of that. And he can say he saw this and he saw that, but he didn't really see it. He didn't smell it. The smell is the most terrible, that odor of burned flesh, you can't describe that, rotting flesh.

And some of that I was responsible for, I mean I directed fire in there. That doesn't help either, you know, when you know you're a party to that. Then you go down and you see stuff like that, you think, what the hell kind of a man am I? It gives you a bad feeling.

The Ones You Hear

Dess Tibbitts

Dess Tibbitts, of Orland, California, started out in the horse cavalry and became a driver in the 712th Tank Battalion.

I always thought the tankers were better off than the infantry, because them poor infantry, I'll tell you, I sat and watched one time there in the hedgerows, watched these infantry dig *allllll* morning, they're digging this big hole right there alongside of the tank. I got out to go to the bathroom, and I hadn't even got my pants pulled up when I heard these shells coming. Well, you don't hear the one that's gonna hit you. But I made it back to the tank, and got in and had my hatch down and an arm fell right across the front of my tank. It had hit right on those guys that were digging them holes. And Hank Schneider didn't know it, but Hank got to crying, and I said, "What's the matter, Hank?"

He said, "Your arm's laying out there. You got hit."

I said, "No, Hank. I had the hatch closed when that thing came. I didn't get hit." I said, "That's one of them infantry's arms." And he was a-crying up a storm. You don't hear that shell that's gonna hit you, but you hear the ones that's just going past you. I heard them shells a-coming, and I run to the tank I didn't even have my pants buttoned up and I made it back in and got the hatch pulled down, just in time.

What Happens to Combat Veterans

Clarence Rosen

Clarence Rosen was a gunner in the 712th Tank Battalion. He took part in the battle for Hill 122, which is the subject of "They were all young kids," but I never got to interview him for the book. He passed away in 1960. This essay, which he wrote before his death, was sent to me by his sister, Viola Oelrich, of Ogilvie, Minnesota.

Let's say you were there! You are moving along cautiously, your eyes straining to detect movement that may indicate an enemy ambush.

You are the tank commander standing in the turret of an M4A4 American tank. The supporting infantry are walking close behind in cover of the tanks. You come to a curve or bend in the road, you caution your driver to slow, you direct your gunner to traverse the gun turret to cover the bend in the road, and there is your objective. A river, a bridge, not a shot fired and you are only two yards from completing your mission. You were briefed by your platoon commander, before starting, that reports from intelligence had it that over 100 Heinie fanatics were to fight to the finish to hold a village, named Susisce, in Czechoslovakia. You again order your driver to move but to approach the bridge cautiously. You spot movement through your binoculars, and as you brace and prepare yourself for battle, to your surprise you discover that they are not Germans, but your allies, the Russians. After handshakes and comments and trying the hand sign language, you radio your commander that the mission is complete, and you learn that this is the last combat mission (in the ETO) you have to accomplish. The War is over. The hasty, bloody, smeary messy war is over after almost a year in combat.

So you come home, the old familiar faces are there, but they are empty. There is no expression in them. They don't know - they haven't been there. They can never know nor can you ever tell them, because they can never understand.

So you're home - you try to pick up the threads but your hands are palsied, and your brain is numb. Somehow in spite of all your trying, the only real thing is noise and thunder and death and hell.

It is quiet now - but soon your numbed mind has you outside into flame and crashing death, you drift back to the time on the Normandy beach, the day you were committed to battle, the terrible roar of enemy artillery shells exploding, the rat-a-tat of machine gun fire, the cries of dying soldiers, laying there with guts some with lungs blown out. The smell of dead Germans and Americans and cattle that were grazing about in line of battle now dead and bloated. And the days you were spearheading when enemy antitank guns sent armor piercing steel through your armored tanks killing the crewmen, wounding the others and in an instant the whole machine burst into flames and you drag yourself out half dazed, then they open up on you at close range with machine guns and rifle fire. You dive head first into a ditch where they pin you down where you have to lay and hear the blood-chilling cries of your wounded buddy burning to death, whose cries are stopped when the ammunition magazine inside the tank blows and now mortar shells are bursting close, and the heat drives you back, in spite of the danger, but somehow you manage to get back to your own lines, to again in a few hours replace and reorganize another tank crew and back to resuming battle. There is no stopping. Numbed and dazed like walking in your sleep, you must keep going.

Now you can thrust an arm into a flame and it comes back seared and blackened. Later come the scars. Contact with boiling water leaves tortured reddened flesh, blisters, and again scars.

These things you can see and comprehend. You understand them because you have had some experiences with burns; therefore to you a burned person is an object of understanding and compassion.

But have you ever been shot at even once in your life? Have you been shot at day after day, week after week, month after month for up to a year?

Have you ever in your life, when driving past a cornfield such as in Iowa or throughout the corn belt, stopped and gazed in wonderment at long wide endless rows of corn? Or have you ever walked out on a high mountain bluff and stared down at the breathtaking depth below?

While on a tour of duty in France, I learned from a source, of the sepulcher of one of my buddies killed at my side in action, and as I was passing close by, I stopped to pay my last respects. As I entered and walked through the gate, there were large letterheads stating that no photography was to be permitted inside. This was a temporary military cemetery, and there, larger than any cornfield that I have ever seen, were row upon row, acre after acre, of white crosses, all in neat, well-kept mounds. As I searched for a certain section, I could not help in my bewilderment to snap photos nor had I up to this time fully realized all the supreme sacrifices that had been made in such a short time and that the cost of liberty and freedom can never be totaled in dollars and cents.

A Letter to Myself

Dale Albee

After 11 months in combat, during which time he received a battlefield commission and three Purple Hearts, Albee put some of his thoughts on paper as he wrote a letter to himself. He kept it hidden for more than 50 years, but revisited it at the urging of his brother Jake.

Sometime in 1945

How hard it is to share with someone who was not there the joy of knowing you were going home but at the same time the deep sadness that you had. The sadness of leaving friends who had shared the toughest, meanest, and most terrible years of our lives. Ones who had protected you and you them in the deadly game of life or death.

To know that suddenly the world you had known of fear – loneliness – and that you could live or die all too quickly in one split second was now to be left behind. But ahead what? We had been away so long. So used to another way of life. We had become animals – now we had to convert back to decent, lovable, caring human beings again. Would people understand and give us help and above all time to reconvert ourselves? One minute we were trained killers and suddenly we were transported to another world. It had taken long hours, and days, for us to become full of hatred for our enemy and to learn to kill without feeling or remorse. I must train myself away from sights – sounds – smells. A red flash of light from a hillside, down the road or in a village meant enemy fire – how would I react to police and fire engines when I first saw their red light or saw even a sudden burst of light? Isn't it strange to be a person that thought he had known all types of fear and now to be faced with one that eats at your heart even though those wonderful words have finally come? You are going home. Lord give me the strength to cope with this new and unexpected fear. To one who had resigned himself to death – with little hope of making it all the way without death or horrible mutilation was almost impossible for me to hope for. Now alive and with orders to return home I am so scared I could cry. Can I convert back now from a killing machine to the decent, loving and caring person my wonderful parents taught me to be? Please, Please Lord – help me once again.

A Difficult Choice

Eugene Sand

Eugene Sand, of Syracuse, Nebraska., drove a recovery unit in the 712th Tank Battalion.

You've heard the guys say about having nightmares and stuff like that. I've never had any nightmares. I probably do a little more reminiscing now than I did for years and years after the war, and now I'll find myself talking about the Army and the war to people that aren't really much interested in it, and I say, "Oh, well, I'm wasting my time with this guy." But there's one thing that stuck in my mind, the first guy I ever saw get hit.

As we were going with the recovery unit, we were supposed to go up for some reason, now that part I have forgot, but there were some infantry guys out there ahead of us. We were supposed to go through the infantry and pick up something. But an infantry guy said, "If I were you I wouldn't be going up there." There was a regular Caterpillar bulldozer just ahead of us, it was blown to hell, and whatever got it was still up there. So we backed off to the side of the road so that we weren't in plain view down that road, and we got out of the recovery unit and tried to discuss what we should do.

I stayed close to the side of the recovery unit. We had cover on one side that way, and when I got to the rear of the tank, instead of just walking on past I looked around. There was one of these old hedgerows on the other side. There was no dirt bank; it was just a brush hedgerow that had been trimmed about shoulder high, and some kids from the Signal Corps were unrolling telephone wire just on the other side of that hedge fence. And probably as far as from here to that van that is setting out there, maybe not quite that far, a damn mortar shell landed right between the legs of one of them. It blew him up in the air, blew him over that fence and he landed in the road. He landed face down, and he was in such misery, I never heard anybody scream like that before. He was beating the ground with his elbows and his knees and he would raise his body up, and he was almost kind of making a circle.

There were just a couple of infantry guys around there and us, and the thought that went through my mind was that somebody should just as well shoot him and put him out of his misery. This guy was in misery, you can't imagine. And I waited probably thirty seconds to see where the next mortar round goes, and here this jeep comes buzzing up there, two medics jumped out of it, and we laid this guy on a stretcher, threw him on the back of that jeep, one guy got back in to drive and the other one was strapping this guy down and giving him a morphine shot, and they got out of there in a hurry.

Many times I've said, "Boy, I never was injured but I still was glad to see those medics show up when they did." I've always wondered, would I have shot that guy or wouldn't I?

Norma

Jim Cary

Captain James Cary was a company commander in the 712th Tank Battalion. He was wounded in Normandy, and again during the Battle of the Bulge.

I had this picture, sort of a glamorous photo of my wife, in my duffel bag which was left behind when I was wounded.

It was taken by a Louisville, Kentucky, studio – she's from Louisville – and all it had on it was, "Love, Norma."

We didn't know that there had been something special about this picture, but I came back and then I was wounded a second time up in the Bulge, and I was back in Madigan General Hospital and people started sending us clippings of a story that had appeared in the Louisville Courier-Journal.

It said that here was this photograph of Norma. Then it said that this guy named Corporal Tornaire who was from Kentucky worked in a quartermaster corps and drove a truck in the 90th Infantry Division. He said he was sick and tired of hearing guys from Texas and Oklahoma, particularly from Texas, talk about how good-looking the Texas girls were, and he was going through my stuff after I was wounded to take out the government property and he came across this photograph. So he says, "Look, this is proof that Kentucky girls are better-looking than Texas girls." It had the name of the Louisville studio on the back. So he took my damn photo, which he had no damn business to, put it in his truck, and started driving across France with it.

So the truck gets bombed and he's not hurt, and he decides that this photograph is a lucky talisman of some kind. The war ends, and he gets more and more curious, he wants to know who is this girl Norma from Louisville. He was from Louisville, too. So he sent the photo and a letter to the Courier-Journal, and they traced the photo back to Norma's family, and they wrote this story about this Corporal Tornaire. The article ended by saying he wanted to meet this girl, but alas, she has a husband.

299th Combat Engineer Battalion veterans with author Aaron Elson, from left: Sam Trinca, Tony DeTomaso, Bill Secaur, Jim Burke, the author, and Chuck Hurlbut.

6 D-Day Veterans Talking "Saving Private Ryan"

Sometimes in an interview or conversation I'll get what might be called a little bit of ancillary history: one historical episode, that is, connecting with another.

For instance, when I was transcribing my interview with James Finn, a veteran of the 294th Combat Engineer Battalion, I was slightly agitated by the fact that Jim's wife was in the next room talking on the phone, which I hadn't noticed during the interview. And when she wasn't talking on the phone, she was either listening to the radio or watching TV.

Then, as I was transcribing, during a lull in the interview I could hear more clearly the sounds coming from the next room, and the TV or radio was broadcasting a basketball game. Not only that, it was a New York Knicks playoff game. One minute while transcribing I was frustrated as heck, and the next, the Knicks were in the playoffs, flashing me back to my days as a copy editor in the sports department of the New York Post.

Or the time I was visiting Forrest Dixon at his home in Munith, Mich., around the time of his 80th birthday, and all over the television was the scene of a fallen Ranger being dragged through the streets of Mogadishu. This upset Forrest terribly. And just by happenstance it was around the first anniversary of the 9/11 attacks that I interviewed Nick Paciullo, a 4th Marine Division veteran of Iwo Jima, and he spoke at length about the episode of post traumatic stress that anniversary brought about.

When I interviewed a group of veterans of the 299th Combat Engineer Battalion in Ithaca, N.Y., in 1998, coincidentally "Saving Private Ryan" had recently been released. Sure enough, the subject came up in conversation.

Jim Burke: The actual D-Day scene was noisy, that's all I can remember. I don't remember anybody saying anything I could hear.

Chuck Hurlbut: He tried to make it ("Saving Private Ryan") that way because that's the way it was.

Jim Burke: That's what I'm saying, it's pretty typical. That's what was so authentic about it, that's all.

Chuck Hurlbut: The first half-hour, if anybody asks me I say the first half-hour was true.

Tony DeTomaso: But the rest was all Hollywood after that.

Chuck Hurlbut: I was so sad he got killed. I thought he's a good egg. If it was a Hollywood film, he would have come home a hero.

Jim Burke: Yeah, but did you notice, he had to put his captain's bars on his helmet. That's Hollywood too. Most guys took those emblems off real quick.

Chuck Hurlbut: I remember seeing a lot of them.

Jim Burke: Before the invasion or after? During the invasion you didn't. They took them off.

Bill Secaur: You know that Ryan, you remember, his cousin is from Syracuse.

Jim Burke: Tommy Niland. We had him at our 50th anniversary.

Bill Secaur: That's right. From Syracuse.

Jim Burke: He was in the 101st Airborne and was a cousin of the real life Ryan.

Jim DePalma: The real family was from Buffalo, right?

Jim Burke: Tonawanda. And this guy was too. Tommy Niland was basketball coach at LeMoyne College for 25 years, maybe.

Chuck Hurlbut: I thought it was so impressive, what really got me, it opens up in the cemetery with the old guy, then at the end, he becomes...

Jim Burke: That was pretty clever.

Chuck Hurlbut: This is Private Ryan as an old man. Then at the end you saw the young guy. Now tell me, did any of you guys ever hear Fubar?

Jim Burke: No.

Chuck Hurlbut: Fouled up beyond all recognition. They used it in the movie.

Jim Burke: Maybe the Rangers had it.

Chuck Hurlbut: I can't find a guy that ever heard it. I never heard it. We had snafu. You heard that every ten minutes.

Tony DeTomaso: What was the name, fubar?

Jim Burke: F-U-B-A-R

Tony DeTomaso: Yeah, but that's just the initials. Well, you know what they were saying, don't you? He just told you.

Chuck Hurlbut: Fouled up beyond all recognition. Like a snafu. Situation normal all fouled up.

Jim Burke: That's why they had that little clerk guy running around trying to find, he's an interpreter for German, he says "I never heard a German word like fubar."

Chuck Hurlbut: I never heard it.

Jim Burke: Nobody did. It might have been a Ranger thing.

Chuck Hurlbut: Now another thing about it. One of the guys gets hit. He's laying out there and they're all putting sulfa on him. Then they take the canteens and they wash it all off.

Jim Burke: Getting the blood off.

Chuck Hurlbut: Why wouldn't that wash the sulfa off? I couldn't understand that.

Jim Burke: It probably did.

Chuck Hurlbut: Every five minutes, washing him off and putting sulfa. I can't understand that.

Jim Burke: It was dramatic, wasn't it? That's the idea.

Chuck Hurlbut: Yeah, but it seems so foolish, put the sulfa then wash it.

Jim Burke: Yeah, but when you're trying to simulate combat it's all screwed up.

Chuck Hurlbut: I guess.

Jim Burke: Everybody's excited.

Bill Secaur: Actually, what we took off the boat, it never showed them taking little boats off the landing barges either in that movie.

Chuck Hurlbut: Well, they couldn't do everything.

Jim Burke: You didn't see any engineers, though, either. Did you notice, those obstacles were up in the water pretty good? This is after the tide's come in. Quite a bit later. this is the third or fourth wave.

Bill Secaur: You know what I can't understand? Why they give the combat infantry badge just to the infantry. You know, we did a lot of infantry work.

Chuck Hurlbut: We should have that. Or its equivalent. The medics got it.

Aaron Elson: I can't imagine the combat engineers not getting combat infantry badges.

Jim Burke: I don't think we should, though.

Tony DeTomaso: Why?

Jim Burke: You can't compare your little infantry stint with a guy who's been there for 30 days in a ...

Tony DeTomaso: We didn't get the credit we should be getting.

Jim Burke: You don't deserve a combat infantry badge either. That's their thing.

Chuck Hurlbut: When the infantry guy in the rear echelon back in Paris gets it ...

Jim Burke: That's their problem.

Chuck Hurlbut: I want it.

Tony DeTomaso: Look at the glider pilots. They didn't get no jump ...

Jim Burke: No, but they got a pin. There's a combat engineer's badge, probably. I don't know.

Chuck Hurlbut: There should be. There is none to my knowledge, but there should be.

Tony DeTomaso: We did a lot of infantry work.

Jim Burke: What, a day, two days?

Bill Secaur: Regardless, you've got to build bridges and be the last one out and blow them up, too.

Chuck Hurlbut: How about the Prum River? We're down putting the damn footbridge, the infantry's lolling up on the bank waiting for us. They're all getting the combat infantry badge. I'm not.

Jim Burke: No they're not, not for that. Thirty days on the line. You talk to the guys up there at Aachen, when they can't talk straight, their eyes are bugging out of their head, they're so damn tired and disgusted, they're the guys that deserve the combat infantry badge.

Tony DeTomaso: They deserve it, no question about it, but so do we.

Jim Burke: You know what happened to us up at Aachen? They were posting signs, one mile, two miles, six miles. All of a sudden these guys stopped us. They were the 1st Division. That's their roadblock. They said "You can't go in there."

I said "Why?"

They said "We haven't taken it yet." And we've got signs, we're supposed to go down in the middle of town and then start out.

Bill Secaur: That movie with Telly Savalas where he was a tank guy, the guy was dressed like an American soldier and he was German, and all of a sudden they asked him a question, and right away Telly Savalas, he knew darn well the guy was lying. And he changed the signs around.

Jim Burke: They said "You can go in there if you want to, but until they come back, we don't know what's down there. And those guys were excited, those infantry men are terribly distraught. I couldn't get any sense of them.

Tony DeTomaso: Most of them are replacements.

Jim Burke: And they're just like the cartoons, grizzly and dirty and they haven't been anyplace except in that friggin line, and they're sick. They're really sick. It's a wonder they don't shoot each other.

***

Sam Trinca

Sam Trinca: C Company got hit bad. I was on Omaha. As fast as that boat hits the beach, as far as it can come in, that ramp will drop. Naturally you're a sitting duck. They're shooting at us, whoever gets hit, I don't know whoever got hit. My part, I jumped off the side.

Aaron Elson: Why did you jump off the side instead of going out the front?

Sam Trinca: Because the guys were coming off so fast, I figured the hell with it, I ain't gonna wait, like you see cattle coming out, a stampede. I was in the back on the boat. Well shoot, by the time they come out, I ain't gonna sit there, I jumped over the side. That's where the water was over my head. You know, the boat is so long; the ramp is in a part where it's a little lower.

Aaron Elson: Where did you go once you got onto the beach? What did you do?

Sam Trinca: First thing we did, we took off our bangalore torpedoes and started fusing them up together, with the rest of the fellows, whoever was alive, start putting them together and we blew up the obstacles, and we made a 50-foot gash.

Aaron Elson: You succeeded in doing that?

Sam Trinca: Yes. We made our gash, but whoever was getting hit, got hit.

Aaron Elson: Who did you see get hit?

Sam Trinca: DeAngelis and Spinelli. They were supposed to be getting on those rubber rafts with the reserves, they're pulled by the boat. They jump off the boat and get some rope. They got hit. In fact I went to school with one or two. We went to school with each other. We played ball. We set up pins in the bowling alley, do anything together. And they got hit. After we made our beach landing, our job was to go in as fast as possible, start probing for mines, which we did. We groped around. And we put tape, to make a path.

Aaron Elson: Did you find any mines?

Sam Trinca: Oh, yeah. We found mines. We pulled them out. And we taped a path so the infantry could come in, they're following the paths.

Aaron Elson: What do you think, when you're probing for mines and you're being shot at?

Sam Trinca: You don't think. You do your job. You know guys are getting hit. If you get hit you get hit. You can't stop and think. You're doing it, that's it. At the same time, we had to face those cliffs, on Omaha Beach. The Rangers were climbing up.

Aaron Elson: Did you see the Rangers?

Sam Trinca: I saw Rangers going up the ladders. And at the same time, we were right out in front of those cliffs where the Germans opened up with flamethrowers. They had pipes. It shoots gasoline and then they light. The pipes were all ready. They had plenty of time to do that. The pipes were throwing out flames, and at the same time they were shooting those fellows down.

Aaron Elson: Did anyone get hit with the flames?

Sam Trinca: I've seen guys get hit. I don't know who they are, I can't tell you. That was a bad thing, very bad. See, before we made the invasion, in England, we got briefed, in a big auditorium, maybe about two or three days before the invasion, and they showed a big screen, the moves of the Germans. They probably had secret agents, they probably were in France or Germany years before, who can speak fluent German, get information and taking pictures. Your life is in danger, those guys there, secret agents everywhere. They brought back to England what they had picked up. Naturally they made the film, and they showed it to us for the invasion. And they showed us the pictures, and it showed the Germans working on the beaches. They had horses pulling those big steel rails and they were sticking them in the sand, making the tripods. Putting the mines in, sharpening them up. And it shows them working, where the cliffs are, it shows the bunkers, where the positions are, where there's any observation towers or anything you can go by, landmarks. They showed all that. They said now this is where this company is gonna hit, this is where this platoon is gonna hit. Gold, Juno, and so forth. We were Omaha. Omaha was our beach.

Aaron Elson: Did you land where you were supposed to land?

Sam Trinca: Very close. Because you had the landmark which showed us. The landmark, according to the picture, was a red, like a steeple or something, a red house, some kind of red structure. That was the landmark we went by. If you see that you know you're in that area. And sure enough, it was there. C Company. So we landed in the right spot.

We came in with decontaminated clothing, in case they used gas. We had so many clothes on, because you had to be prepared for anything. You had your fatigues underneath. You had your decontaminated clothes on top. You had your pack full of ammunition and explosives. You've got your two bangalore torpedoes. You had your bandoliers full of ammunition. You had your rifle. Nine pounds of M-1. Man, you realize you had ninety to a hundred pounds on your back.

Aaron Elson: Did you see the tanks on the beach?

Sam Trinca: No, the tanks didn't come in yet. The tanks couldn't come in until we removed the obstacles. Armor came in as we were clearing up the beach. We had to clear up whatever's ahead. The combat engineers first, anything in the way we had to do it. They gave us some support, even if they were on board ships or on boats, they were shooting. Keeping the enemy down as much as possible.

Aaron Elson: What did you see on the beach? Did you see bodies?

Sam Trinca: Oh, yes. I saw many bodies. Floating in the water. bodies all around. Guys up, begging for help. But you don't turn back. You can't turn back. There were medics, they help out the best they can do, and there's the landing crafts that they use; if there's any landing crafts still coming in, if they can throw any bodies in the boats, fine, all well and done, or anybody that's hurt, you just get them and throw them in as fast as you can grab them, and they turn around and take off.

Aaron Elson: What images stand out that you remember specifically seeing?

Sam Trinca: Just the images that you see in your mind just that day, where they pick them up and try and do the best they can. Get the bodies or whatever, whoever's wounded, try to pick them up as fast as they can and try to help them out. But not us guys.

Aaron Elson: Were you scared?

Sam Trinca: Scared? Oh, you didn't read the article. I had an article in the paper. One of the guys, he was asking about, what was that movie?

Aaron Elson: "Saving Private Ryan"?

Sam Trinca: And they called me up because they knew I was one of them. They said "Mr. Trinca, you're one of the boys who was at Omaha Beach. Could we have a little information about it? What we want to know is, does the producer hit the spot?" They wanted to know if they made that movie as close as they said. They claimed that that movie was very close to the real invasion.

Aaron Elson: Was it?

Sam Trinca: I don't know. So they asked me that question. "Did you see the movie by any chance?"

I said "No, I didn't."

"Oh, well," he says.

"I said, "Give me a ticket, I'll go and see it." How could I prove what the movie looks like if I haven't seen it? I said, "I can tell you my version of the story." So I told him my story, like I'm telling you now.

He said, "That sounds pretty close to it."

I said, "I can't tell you what the movie is."

He said, "Are you going to go see it?"

I said, "I don't know, maybe yes and maybe no." But I didn't see it. They were trying to compare how close they came to making it real. Because they said a lot of it was a lot of bull. But some people thought it was real.

Aaron Elson: Going back to scared.

Sam Trinca: Oh, scared. I was afraid to even mention it in the newspaper. I told him, "You know, I was so damn scared I wet my pants so many times it wasn't even funny." I mean a lot of my friends downtown, they said, "You're that guy who wet his pants, aren't you?"

I said, "I'm not ashamed to say it. Yes, I did wet my pants." I tell my friends. In fact, I was at a store that my wife goes to, and I saw a lady, a friend of ours. She said, "Hey Sammy. I read the story in the paper. So you wet your pants, huh?"

I said, "Yes I did." I said, "I'm not ashamed to tell you."

She said, "I don't blame you. I'll bet you some guys did more than that."

I said, "They probably did." Then I said, "Thank god I was in the water anyway, I was soaking wet, so who could tell the difference?" She started laughing. What the hell, I was soaking wet anyway. How many times did I have to wet my pants? I don't know, from there to there? Hell, yeah, you bet your life I was scared. Scared? Like I said, when you jump off that boat, you don't know what you'll do. You know you've got to get the mines, your job is to do what you're supposed to do and keep on going. You're here. You watch yourself, you try and help your buddy if you can.

Aaron Elson: One thing in the movie, at the very end, they show some dead fish on the beach.

Sam Trinca: Fish? I don't remember.

Aaron Elson: Maybe he did that for effect.

Sam Trinca: I could see that. Why shouldn't there be any dead fish? With all those shells going in around there? That's just holy Hell. All it is is just Hell. All you hear is bang bang bang bang bang ding ding.

Aaron Elson: Did you ever get to a point, especially after D-Day, where you said, like a fatalist, that you thought that you were not going to come home?

Sam Trinca: Yes. Well, many times I figured, I've seen my buddies get hit both sides. I never knew when my time was gonna be. Yeah, many times I figured "This is gonna be it." You know, as you keep going along you stop and try and rest, and you're thinking, what's up ahead? And you're trying to drag what you've got with you. I had a machine gun with me. I can't remember whether Chuck helped me carry the tripod. Someone did, anyway. We had to be close together. In fact I've got pictures of Chuck with my machine gun. He's probably got pictures, the same ones too. All I know is you just keep moving ahead. After the beach, as we got in just a little further, another obstacle was the hedges. Hedgerow after hedgerow after hedgerow, they were a sonofabitch. Those hedgerows, even a tank couldn't penetrate through. That's how those farmers had them built.

Aaron Elson: What was your rank?

Sam Trinca: Pfc. Private first class. You know, it's hard, like I told the fellow in the newspaper. I said "Look, what I'm saying, if you weren't there, you can imagine what it was like, but you could never describe it. Never." And he put that in the article, he underlined it in the paper when he wrote it. You could never describe it. And it's true. That's why he was trying to tell me, like a movie? How in the hell can I tell you what the movie was like if I didn't see it, and then I could tell you about it. But then, like, even my wife says, "Why don't you tell them to give you a pass to go see the movie? They can afford it." They never did.

Write Your Story

Jerome Auman

On Memorial Day of 1998, I attended the dedication of a new library at the Eldred World War II Museum in Eldred, Pa. The museum director, Kurt Pfaff, invited some area veterans for me to interview. One of them was Jerome Auman, a Marine veteran of World War II.

We had a memorial service for the 22nd Marines in Canton, Ohio, this year. Our chaplain was a Marine in World War II with the 22nd. He wasn't a chaplain in the service. He was in the 22nd Marines, and after he came home he went into the ministry, and he was in charge of the memorial service in Canton for the seven of our fellow Marines who are no longer with us since last year. The services at the First Methodist Church there were supposed to be at 10:30 in the morning, and we were there at 10 after 10. It's too early to start the services, but it looks like all of us are here. So the chaplain asks, "Is there anybody who wants to say anything to the rest of the people to use up a little time?" He didn't want to start the service ahead of time because somebody might still come in.

I was sitting in the first row. I look around. Nobody seems to be making a move. So I put up my hand. Reverend Mack says, "Yes, Jerome, what is it?"

I stand up and turn around and face the crowd. He says, "Oh, no, Jerome. You've got to come up here to the pulpit and use the microphone so everybody can hear what you have to say."

So I go up to the pulpit and I look around. I said, "Marines. Ladies and gentlemen, we're gathered today to pay homage to the seven fallen comrades who are no longer with us this year. And I just wonder how many of the seven wrote their stories, because if they didn't write their story, nobody can write it for them. Only they could write it. Now how many of you have written your story? If you don't write them, there's nobody can write it. Look at our age. We aren't going to be around, as seven are showing us that are no longer with us. Write your story. Your children, your grandchildren, and history needs these stories. I'll make a copy of mine available in the hospitality room so that it will give you an idea of how to do it. Now I know a lot of veterans never want to talk about their experiences, but that's not the attitude to have. You should want to write this down, want to record it. Maybe you weren't ... I didn't do anything to merit a Congressional Medal of Honor or anything like that, but I've written my story. There'll be a copy of it available in the hospitality room when we get back. That's all I want to say except one thing. We'll never know how soon it'll be too late."

Well, that took a few minutes. Somebody had to break the ice. Then a couple of other fellows had something to say about their units. Of course, it got to being 10:30 and we went on with our memorials, which lasted probably about a half hour, and like I said, I was sitting in the front row and I was one of the last to leave. Reverend Mack said there's gonna be coffee and donuts in church, you're welcome to help yourself and visit with the rest. So after the service is over, everybody's going down the aisle. I was about halfway down the aisle, and this other Marine came up and put his hand on my shoulder. I never saw a man cry like this man was crying. The tears were actually dripping off his chin. What do you do when you see somebody crying like that? He put his hand on my shoulder and he said, "Mr. Auman, I wish you would have told me this a few years ago. I lost my son – he had cancer – about six months ago. The night before he died, he looked up at me and said, "Dad, you never once told me one thing you did as a Marine. I wanted to know so bad. It's too late now." That man cried. It actually makes tears come to my eyes just telling the story. And he said, "When I get back, I want to write my story. I have other children."

Put yourself in that position.

And then another story, on a more brighter side, coming out of the same thing. We get back to the hospitality room, and Vern Hodges is his name, was reading my story. And almost from the first word, "Hey, I crossed the country like this. I went in this ship that left. I went to this heavy machine gun thing. I got in the MPs. I came back to Pearl Harbor. I went under the Golden Gate Bridge. I got this leave. I did this. I did that!" I had just a little wee picture, about yea big, of our Marine company in Samoa. I said, "Well, can you find yourself on this picture?"

"No," he said. "I can't."

I said, "Well, what I'll do when I get home, I'll have copies made of it, and I'll get 8 by 10s, and I'll send it to you." And I did. And we've become real close friends.

The Eldred World War II Museum is located at 201 Main Street, PO Box 273, Eldred, PA 16731. For more information about the museum and library, call (814) 225-2220.

Frenchy Belleperche, 2nd from left (photo courtesy of Paul Belleperche)

Frenchy

Jerome Auman

Sometime around Christmas of 1942 I was sent to British Samoa. American Samoa and British Samoa were on the front line – the Japs hadn't gotten any farther than that – so we were the front-line defense. I was transferred into what they called a heavy .30-caliber water cooled machine gun battalion. And we trained. Oh, man, what work! We had to carry them uphill. The tripod and the gun, the water tank and two boxes of ammunition were on a cart. The cart had wheels that look like bicycle wheels, they're about 26 inches; a bicycle wheel has real fine spokes. You had a tongue on it, and I think they used us like mules to haul this cart. Two guys pulled it. And when you got to where you couldn't pull it any more, they had a rope to put on front and two more could help pull it. And when that got to where you couldn't pull it any more, you'd carry it.

In that squad, there were 14 men. The first man carried the tripod, the second the gun, the third the water tank and a box of ammunition, and each one after that carried two boxes of ammunition. I don't think I ever worked so hard in my life.

That went on for about six months. Then we left British Samoa and went to American Samoa, and from American Samoa, we were going to board troop transports for the invasion of Tarawa.

The night before I was to ship out, my name was called to report to the first sergeant. So I trotted down to the hall. We had been sitting watching a movie – they hung up a big sheet in the field and they projected a movie on the white sheet, and you could see it from either side of the sheet. They had loudspeakers set up for the sound. So I went to see the first sergeant – they didn't have an office; they had a table sitting out under the trees. And he said, "Pack your bag. You're being transferred."

What a relief to get out of work like that! I was transferred into the military police company on the island. This military police company had the brig for the whole South Pacific area. If you got in trouble, you wound up in our brig. And I spent about six months in the military police.

There was a Navy seabee company on the island, and they were in charge of the generator for the whole island. So it was a pretty big generating station. And they also had a submarine base there.

Each MP was assigned two fellows from the seabee company to take a month's tour of the MPs, because there were sailors and Marines coming in, and they didn't want the Marines picking on the sailors.

There was this one fellow by the name of Frenchy Belleperche. I don't know what his right name was but Belleperche was his last name. He was from right across from Detroit, Windsor, Ontario. He was a Canadian, but he enlisted in the United States Navy and became a seabee.

Frenchy and I became very close friends. And one day, he came to me and said, "Jerome, I'm in real big trouble."

I said, "What did you do, Frenchy?"

He said, "Well, you heard about the stink being raised on the island. I'm going to be arrested tomorrow. Will you do me a favor?"

I said, "What is it?"

He said, "Take this cigar box."

"What's in it?"

"Look."

I opened it up. There was $13,800 in $20 bills.

I said, "What am I gonna do with this?"

He said, "Take it. You're going back to the States pretty soon. Take it home."

I said, "Frenchy, how in the hell would I explain this much money making $21 a month?"

He said, "Oh, just take it!"

I took the box, and I put it in my footlocker, right down on the bottom, and put everything on top of it. Nobody else knew anything about this.

Well, the case came up. What they had been doing was taking what they called torpedo juice – that's the fuel that you shoot the torpedoes with – and they were distilling it, making 100 proof alcohol. They were selling that for $100 a gallon, and they were supplying every ship and anybody that wanted alcohol on the whole island. Frenchy's share was over $13,000. He said, "I can't use it. I'm going to be arrested. They're going to find it."

What had happened was that two fellows found out about the making of this liquor and they wanted to cut in on it. And the guys who were making it said, "No. We don't want anybody else, because we're a pretty nice, well-regulated organization." So instead of leaving well enough alone, these other two Navy fellows decided they're going to be smart and they're going to make their own. But what they did, they took paint thinner, and sold it out of the barrels just like that. They didn't distill or change it or anything. I don't remember the figures, but three or four men died. Some went blind. Some went crazy. When you get on an island and that kind of stuff turns up, it is a stink. So the roof caved in on Frenchy's operation.

I kept the money in my footlocker for a couple of months. Frenchy got court-martialed. I don't know what rank he held but he was busted down to the lowest rank in the Navy, and he got six months in the brig. Well, if you knew the guy, he was one hell of a nice man. And I'm not ashamed to say he was a friend of mine. Almost from the first day he was in the brig, they made him a model prisoner, and he only had to sleep in the brig at night. In the daytime, he was in the officers' mess doing cooking and cleaning.

In the meantime, I was going to go back to the States. At any time, no general will travel alone. If a Marine general was ordered back to Washington, he had to have a detachment of Marines travel with him. I had 16 months overseas at that time, and I got chosen as one of the Marines who were in this detachment to travel back to the States. So I went to Frenchy and said, "Frenchy, I can't take this home." On the inter-island boats there's always three or four MPs inspecting baggage. And what are they gonna do when I get back to San Diego or San Francisco? How will I explain that?"

He said, "I don't know, but I want you to take it home."

I said, "No, I'm not taking it home. Here. It's yours. You're free all day long. You hide it someplace." So I gave him his money back.

And I didn't know it, but when we came home, through customs, there's not a soul who even said, "What's in this bag?" But how would I have explained that lump sum of money on $21 a month?

I was discharged from the Marines in December of 1945, and in 1948 I had to go to the VA in Erie for a physical, because I was discharged with 10 percent disability for arthritis in my hips and knees. I was married by then, so my wife and I went to Erie and I said, "Let's drive out to Windsor and see Frenchy." We went to Windsor and looked him up. We had a good time together for four or five hours. And he said, "You know, I wish you would have never given me that money back because I spent every damn penny of it in poker games."

I never told that story. I've written my story. Kurt has a copy of it here [at the World War II Museum in Eldred, Pa.]. I never wrote stuff like that into it because, well, I figure that's not part of the military thing. It's a story. But I could write a book this thick if I had to write everything that I said or did.

But anyhow, to get back to the general coming back to the United States, on the way back we stopped at Pearl Harbor, and we were going to be there a couple of days. We were allowed shore leave. So I go ashore and look around, and who do I find but my squad leader from the .30-caliber water cooled machine gun, which was the Second Defense Battalion of the 22nd Marines. And I said to him, "Where are the rest of the fellows?"

He looked at me, and he said, "You're looking at the only one that came back alive."

And I don't have to tell anybody that was there how I felt. To think, "Thank God for the transfer to the military police in Samoa." Because I, too, probably would not have come back.

The Bandleader

Stan Freeman

Stan Freeman began his service with the 11th Cavalry at Camp Seeley, California, in 1941.

You know what I remember? I got a weekend pass to Long Beach. There were four of us that were buddies in the outfit and three of us got together, one guy couldn't go, I guess he was on KP or guard; we went out there for the weekend. And on Sunday night, one of the military outfits had a band that played a concert on the pier, and the three of us were walking up there. This was after Pearl Harbor, and we were in full uniform – boots, spurs, the whole bit – and they started playing the Star Spangled Banner. They always ended the program with that. We came to attention, and when you clicked your heels together you made a metallic sound and it was loud. We did that, and I think the 3,000 people turned around and looked at us instead of looking at the band.

There was a Majestic Ballroom in Long Beach, and this was the first time I ever had an experience of running into a major bandleader in person. At that time he wasn't a major bandleader. We went in and the guy at the gate – it was a nickel a dance, or ten cents – there were about five of us, and he gave each of us free tickets, just handed them to us. He said. "Go on in guys. I'm drafted tomorrow."

We went in, and the Stan Kenton band was playing. He wasn't famous then. He was the house band at this ballroom. And the thing I remember about that is he had four trumpets, four trombones and four English horns in his band, and when he let it out, he'd blow you off the dance floor.

A WALK IN THE SUN

After the war was over, I paid a hell of a lot of good money to go listen to concerts by Stan Kenton.

London War Child – 1940

Margaret Hofman

My war started when I was about three months old. My Dad had been called up into the army and my mother had been living in a small apartment in East London. The German planes had already been flying over London and dropping bombs, mainly at night. After one raid, my mother took me in the pram to post a letter to my Dad to tell him that we were all right. While she was away, an undetected land mine exploded and the apartment was just brick dust so she returned to her father's house in the dock area of London.

This was an unfortunate move because the next thing the Germans bombed was the dock. They bombed it by day and night until even the water burned with the contents of the warehouses tumbling into the water. My Grandfather had forbidden all his 'children' (all grown up with families) to go into the large warehouse down the road because he said it was a death trap. It's funny what you do, though, to escape from noise and to get away from the scream of bombs. People crammed into the large warehouse, someone brought in a piano and local teachers and others organized singing to take people's minds off the bombs. The warehouse received a direct hit and over 200 people were killed by the blast. Others died not directly as a result of the bomb hitting them but they were crushed to death by the huge heavy walls of the warehouse.

Apparently we were three days and three nights in an underground shelter in Granddad's garden. Crowded together, the family were not too pleased when the local police knocked on the door and made them take in two more people who had been a mile away from their home when the air raid began, especially as they were none too clean. After the dawn broke and everyone crawled out of the shelter, not only were we covered in dust but covered in fleas from our two uninvited guests!

Granddad said if we stayed where we were we would not survive another round of bombing like that. He had already bought a small plot of land in Essex and built what he called a wooden holiday home there – it was not so much a holiday home, more of a two-room shed! Anyway, the garden was pleasant to look at, which was a good thing as we ended up with eight of us living in the wooden house and the men camped in a tent in the garden.

Eventually we sorted ourselves out. Poor Mum had to live with her mother-in-law, who was hardly a sweet-tempered person. Dad's mother, although born in England, had Irish parents and she was prone to be bad tempered and a bit of a drama queen. She made matters worse by telling Mum all the BAD news of the war, and her prediction of what was going to happen.

Mum moved out and rented a small bungalow which got us away from Nan but made it more difficult for Mum when the air raids started. By this time she had another baby and when the siren went, my Granddad Hickman (kind sweet and gentle – how on earth did he choose my Grandmother for a wife?) would come round and carry me round to the big underground shelter in his house and Mum would carry the baby.

Our shelter was called an Anderson Shelter and was built half in the ground and half on top. The top half had curved corrugated iron sheets and you piled lots of earth on it and hoped for the best. It tended to be damp and if it rained you all had to lift your feet up because the water seeped in from underneath. Grandmother Hickman had chosen the house just after the war started because of the large garden and peaceful neighbourhood; unfortunately she failed to notice it was quite near the railway lines and in a direct line with the Shell Oil Refinery five miles away. Consequently Germans coming in to bomb the oil refinery would miss and others would continue to have a go at the railway. I don't know how old I was when we sat there listening to German planes coming over the shelter on their way to London and we then had to stay there until they they came back. Of course if they missed their target they would jettison their bombs before they were over the Thames Estuary.

My Uncle was home on leave, he was only 19, an aircraft engineer and he was stationed in some quiet backwater. He was fascinated by the planes and my Grandmother was having hysterics, "Come inside the shelter. They will see you." Uncle was a fidget and wanted to see the planes coming back so he made us all a cup of tea and we sat there until the familiar drone came nearer and nearer. Suddenly I realized *he didn't know about the railway lines*. I didn't say anything, everybody else was getting ready with their hands over their ears when suddenly four bombs cascaded onto the railway lines now gleaming in the moonlight. It was so near the whole shelter lifted up and then went down. My Uncle dived in head first in a state of shock and our tea went everywhere.

At one stage we used to go to sleep with our clothes on. It was called a siren suit, kind of like a child's stretch pyjamas so that we could get into the air raid shelter quickly and not get cold.

We had gone to bed at about 8 p.m. – we were cold and coal was rationed and bed was the warmest place. I slept with my mother and she would read in bed. It must have been winter because it was dark – we had thick lined curtains so that the light would not show through and in a raid you just kept the lights off anyway. Suddenly the air raid siren went, I was half asleep but my mother was up and into the back room for my baby sister (I say to her now, I don't know why she chose you first to get into the shelter) .

By the time it was my turn large lumps of jagged shrapnel were clattering on the top of the shelter and the ground, my sister was crying, my mother was frantic to get me, but I didn't care, I was warm and probably tired anyway. Suddenly the railway lines got hit again, the house shook, two windows broke, then my mother came rushing in. In the explosion the vibration had shaken open the wardrobe door, my mother walked straight into it and walked around holding her head. I laughed and laughed, it was like something out of a Charlie Chaplin film. She was so mad, she clouted me round the head and said, "See if you think that's funny."

We rushed into the shelter, she a nervous wreck and nursing her head and my sister screaming. Suddenly I looked up and saw in the sky a long object with fire coming out of the back of it, but really quite close. "Isn't that wonderful?" I said.

"Yes, so wonderful, that is what is going to kill us all if you don't get inside."

The V1 was certainly a marvel of German technology but we all knew that as long as you heard the engine you were safe. Once it cut out, its descent was rapid and it could take out six houses .

My children asked me what our food was like – pathetic is the answer. We grew potatoes and tomatoes, picked berries and swapped things with neighbours. We were rationed so severely that even now I don't eat meat as my ration had to go to my sister who used to be ill with asthma. After the war when we came off ration, I had got used to not eating meat so I never bothered. For a family of three we would have 4 ounces of meat a week – 2 if it had been a difficult time for the supply boats coming in. England as an island depended on overseas help for food. If the U boats were active we all went without as most of the food went to our army. Two eggs a week or else that dreadful egg powder.

I always thought our food rationing was pretty dreadful until I met my husband who was a child in Holland during the war and who really knew what starvation meant.

As for our little family, Dad survived D-Day and went on with the Middlesex Regiment through France, Belgium, Holland and Germany and through great good fortune came back to us. How did the war affect him? Well, he couldn't stand noise, we never had the radio on very loud, he would never listen to a memorial service. He would never go swimming. Years later I learned on D-Day he was dropped too far out and the vehicle he was driving was nearly submerged, he negotiated debris, burning vehicles and bodies to get his crew onto Sword Beach and then off into Normandy.

While his officers were having a meeting near him, mortar fire sped over the trees killing them all in front of his eyes. His confidence never returned, he could never make a decision and was left with stomach ailments that plagued his life.

He looked after us both when my mother was ill immediately after the war and our bedtime stories were Operation Overlord and the time he just avoided a mine in France. We knew all the generals and battle plans. We often spoke of old battles but he would never return to France with me. He died of a heart attack aged 69 but his wartime stories are etched in my mind forever.

Rambling Jack

Ed Stuever

Ed Stuever was drafted in 1941 and assigned to the horse cavalry in California before the 11th Cavalry was sent to Fort Benning, Georgia., and mechanized.

For the first six months that I was in the cavalry, you had to train a horse that was almost from the wild. Some of the horses had only been ridden once, and you had to go out in the corral and train that horse yourself. So I trained this beautiful horse, and the day that they issued the horses to us, I was so proud, I walked up there with that horse.

"Sorry. He's been taken by Officer Clemens."

I said, "What do you mean? That's my horse! I broke him in! I'm going over the hill. The hell with you guys!" Oh, man, I was mad.

The first sergeant, he was a mean old bugger. He put his arms around me, and he said, "I know how brokenhearted you are. How you worked, you trained that horse. There's a few good ones left. You see that one with his ears sticking up there? That's Rambling

Jack. We've had him awhile. He's a good horse on parade. You be good to him, he'll be good to you."

Ohhh, I was disgusted. I went up to him and said, "Come here you little reprobate." I saddled him up and boy, he gave me a mean ride. And I said, "Why, you ain't gonna get the best of me."

I learned that he was mean. If you weren't ready, he'd go back and bite you. He grabbed me on the leg and tore my overalls, but he didn't hurt me. Anytime I went to saddle him up I had my elbow in him. If he went to bite me I'd hit him with my elbow on his cheek, and that's a sensitive spot. "Jack!" I'd always yell at him, "Jack!"

When we'd go out in the corral and we had to get our horses, everybody had to run. I'd say "Jack!" and he'd come to me and I always had oats in my pocket. When we went camping on maneuvers, we used to camp out overnight, and the enemy from the other troop would come and try to steal our horses. So they put me on the outpost because they knew that Jack would bite them, and he did. He got one of them by the arm and he wouldn't let go. And he's screaming bloody murder.

And he was hell on maneuvers. When he met the enemy and they blew the bugle charge, I couldn't hold him. He'd lock that bit in his mouth and he'd take off. He went right through everything. I couldn't hold him back. They're yelling at me, "Control that horse! Control that horse!" You couldn't control that horse.

Then on maneuvers one time we came to a river. Real steep bank. And we had to cross that. There were deer paths and cow paths going down there. It was quite high, maybe 25 feet, and we had to go down to the bottom. So they couldn't get across that. The first sergeant says, "Stuever, come up here with Rambling Jack. You get up there with him, and he's gonna lead us across. I'm gonna be right behind you."

I said, "Ahhh, you'd better be." I knew Rambling Jack loved water. He couldn't wait to get down. He almost left me up there. Then when he gets down there, in the middle of the stream, he lays down in it. He's got my leg under him, my gun went in the water. What a predicament I was in. And you couldn't wear spurs, he didn't like you touching him with spurs. So I didn't have my spurs on, and I gave him a real good kick. Boy, he flew up and he was mad. What a horse that crazy nut was.

The Cat

George Noorigian

George Noorigian of Spring Lake, New Jersey, was a bombardier in the 445th Bomb Group. I interviewed George on March 30, 1999

Usually, when you got a new plane, they wanted you to put a name on it. So when we got the new plane and we were talking about naming it, our crew chief came over to us and said, "Fellows, can I say something?"

I said, "Sure."

He said, "Every time they put a name on a plane, the plane gets shot down. Do me a favor. Don't put any name on this plane." And we never put a name on it. He said, "Every time, the plane gets shot down. Maybe not the next mission but it'll get shot down." And then, every time you flew a mission they would put a bomb on the side of the plane. It showed you how many missions that the plane flew. We had one plane there, they called it the Cat. You should have seen all the bombs on that thing. They couldn't shoot it down. And the crew chiefs, day and night they'd have to work, they always had trouble with that plane. They said, "One of these days it's gonna be shot down and they're gonna give us a new plane."

There was one group that had a bad time. They lost so many planes on the missions that they were going to disband the group and move the guys and the planes to the other groups. They said, "We're gonna give one plane to the 703rd Squadron."

Right away, the fellows with the Cat said, "Ohhh, finally we're gonna get a new plane!"

They told us to go and fly to the base, which wasn't far. You had to be careful when you flew there because they had so many groups there, all close together, when you're in a flight pattern you had to be careful you were not in somebody else's flight pattern.

We were going to fly the Cat up there, leave it, and come back with a new plane. So we went to that base, and we went up to the tower and said, "We're from the 445th. You've got a plane for us?"

"Yeah," the fellow in charge said. "There's the jeep. He'll take you out to it."

So we all got in the jeep and we went out there. And as we're going along we see all these planes. Finally he stops in front of this one plane. I said, "Is that the plane?"

"Yes, that's your plane."

It was another Cat! We looked at it. What a mess that plane was! I said, "We were supposed to get a new plane!"

"I'm sorry. This is what was assigned to you. You've got to take it, fellows. You've got to go back with her."

Well, Jesus Christ, we looked at that goddamn plane, and I'm telling you, when they got the thing ready to go, and they were starting up the engines - they had trouble starting the engines. It was terrible. So the co-pilot, Paul Pouliot, called me in the back. He said, "George, put your parachute on. I don't even think we can get back to our base. We may have to ditch this plane." He said it was terrible.

We managed to take off, and we only had a short distance to go. And we pulled into the base. They were all waiting for us to come back. We came in for a landing, and each plane had its own place to park, so we knew where to go. We pulled the plane down and we went into the parking spot. The crew chief was there and the guys said, "Where's the new plane?"

"This is what they told us to take and that's it."

Well, the guys almost died. They almost died. You know, that plane was shot down after two missions? Can you imagine that?

"Some Poor Bugger"

Patsy J. Giacchi

Pfc. Patsy J. Giacchi was a survivor of Exercise Tiger, the ill-fated practice D-Day landing that is perhaps better known as "Slapton Sands," the area in England that resembled Utah Beach. LST 508, on which Giacchi's quartermaster company was being transported, was torpedoed and sunk in the middle of the night. Giacchi survived more than four hours in the English Channel and recovered sufficiently to take part in the actual invasion of Normandy.

When the war ended, I was in Germany, in a town called Ulm, U-L-M, we say Ulm, but they say "Oohlm," the Germans. Out of the clear blue sky over the loudspeaker they say, "The war has ended! The war has ended!" Here I am in a foxhole talking to one of my buddies. "What did they say?"

"Pat! The war has ended!" You'd see there were some of them out there going crazy. Guys were shooting each other by mistake! GIs, yes, they were shooting themselves, from the excitement. They tried to tell everybody, "Calm down! Be careful!"

And I was in a foxhole down there, "The war is over! The war is over!" I was crying in the foxhole from joy, I couldn't believe it. The following morning they called formation outside, they said, "The following names, please step forward." Finally, "Pfc. Patsy Giacchi!" I step forward.

"Okay," the captain says, "you guys are all going home."

Boom. One guy passes out from the excitement. I couldn't believe it. I was 21 years old then.

The following day, it was a clear blue sky, "Cover up your foxholes." Before you know it, they put us on trucks, the big Army trucks, they load you up in there, hold the back door, the driver takes off, and he's going for miles and miles and miles, before you know it, you're riding almost a whole day. You ride from one part of Germany to another part. You bivouac. Five in the morning you get up, take your tents down, they give you coffee. Boom! We start to travel. Before you know it we're deep into France. Another couple of days we're in Le Havre. And there's the big liberty ships waiting. But before all that they had to give you the inspection, because a lot of guys caught venereal disease. If you had a venereal disease, they took you out of line. They put you in the hospital, and they cured you there, then they would send you home. A lot of guys were pulled out, who had syphilis, the clap. A lot of soldiers caught VD over there. They taught them how to use prophylactics, but a lot of guys were stupid, they got drunk

Now I'm on a liberty ship coming home, I can't believe it. All of a sudden, after about a week, they said, "Here it is, boys." We come around a big bend, and they say, "You guys want to see it? Go up on the top deck, you can see it."

What is it? We're hitting Newport News, Rhode Island [actually, it was Virginia]. And you look, as you come in, you can't believe it, the United States. And a big sign that says, "Welcome home, boys! Well done!" I'm on the deck there, I'm crying, I can't believe it.

All of a sudden you start to line up. I said, "Pat, be careful now, Pat be careful, you made it through the whole war. Don't get killed, don't fall off the ship or get hurt." Careful. Careful. Then they call your name, you step down. So when I got off, I did what they all do: I kissed the ground. I cried.

Then, "All the guys from New Jersey, line up one one side, you're going to Fort Monmouth." All the guys from another state, they line up somewhere else.

So I went to Fort Monmouth. The following morning we had to go to a big church. You had the organist up there playing songs.

They said, "When your name is called, step up, salute the officer who's giving you your discharge papers, make a turn, go back, you're discharged from the Army."

"Pfc. Patsy J. Giacchi."

"Pfc Patsy J. Giacchi!"

A guy goes, "Hey, that's you!"

"Oh, yeah! Yeah!" I go up there, nervous. They give me that paper. I walk back.

"You don't go back and sit down! Get the hell out of here, you're discharged!" I see an officer go by, I salute him. He says, "You're not a soldier anymore."

"I'm sorry, Sir."

"That's okay."

So I'm the only guy that's coming towards Hackensack. They told me where to get the next bus to go from Fort Monmouth to Newark.

I'm on the bus a couple of hours, and then the driver says, "Newark, Penn Station!"

"Penn Station!" I remember that from when I was a kid! Penn Station in Newark.

I got off the bus. I went to a telephone booth. There were many guys there. My time came. I took out my wallet. Now, I didn't know my phone number, because when I left it was three years ago. So I opened up my wallet to look for my phone number. I had my mustering out pay. They gave me three hundred dollars when I got discharged, plus I had another two hundred dollars. That's 1945. That's a lot of money. So I'm nervous, all of a sudden, "Hello, who am I speaking to?"

"You're speaking to Nellie." Now Nellie's one of my sisters.

"Nellie, please, now don't get excited. This is your brother Patsy."

"Who?"

"Patsy!"

"Who?"

"Patsy!"

Boom. She dropped the phone. She passed out.

Jane picks up the phone, my older sister, and she "Who's this?"

"Jane, please, it's your brother Patsy. Please, don't get excited, I'm in Newark, New Jersey."

"Oh my God!" We'll send somebody."

"No, no, no. I'm coming home. I'll take care of it. I'm coming home."

"Patsy, please be careful. Oh my God!"

I left my my wallet there. This is the god's honest truth. I had some change in my pocket, and a couple of bills. I got on the bus. I took the bus from Newark to Hackensack. Then from Hackensack on Main Street I took a taxicab to West Street, where I lived.

As I'm coming around the corner, they've got the thing lit up for me in front of my house, "Welcome home, Patsy!" All my neighbors are waiting for me. I get out of the cab and and they're grabbing me, my mother's trying to grab me, my sisters were there. The neighbors were there. Across the street the DeLorenzos. "Ohhh, Patsy, it's good to see you" and everything else. Who's pulling me here, who's pulling me there.

After about two hours, some of the neighbors disperse, we go inside, we started talking.

Then Jane says, "Gee, Patty, have you got any pictures?"

"I've got one or two pictures. JANE! My wallet! I left it in Newark."

"Oh my God! How much was in it?"

"Five hundred!"

"Five HUNDRED?!!!" Then it was like five thousand.

We call up Newark. They say, "Would you please come down?"

We get down to Newark. I go where I made the telephone call from. Behind the counter, there's a couple of cops there. They said, "Soldier, we get this every day. You're going to have to give us some real good detail. Everybody tells us 'black wallet, brown wallet,' something like that. Tell us if you can what's in your wallet."

"Well," I said, "I've got a couple of this, a couple of that."

"Keep on going." The other guy's writing it down.

I said, "Okay. I've got it. Okay, now look."

I was always excited. "Take your time now," he says.

"Okay. You'll see a picture of my girlfriend, an Italian girl with long black hair. She's got a dress on" – I bought her this dress – "and the dress has got an emblem of a little parrot."

That did the trick. They gave me the wallet, with the money in it and everything else.

"Sir," I said, "who returned this for me?"

He said, "A little old lady. She said, 'Some poor little bugger left his wallet here with all his money. Please see that he gets it.' "

I said, "Can I give her a reward?"

He said, "She doesn't want a reward. Just take care of yourself, she said."

Vern Schmidt, left, and Aaron Elson

Lice

At the 2016 World War II Weekend in Reading, Pennsylvania, a young man who previously had purchased some of my oral history audiobooks remarked that one of the stories was a real eye-opener. It was the story in which John Sweren, a former prisoner of war, remarks that when he thinks back on all the good things that happened to him in his life, and a lot of good things happened, like the births of his children, his daughter's wedding, one of the best days of his life was the day he got deloused. Following are two of the many stories I've heard from World War II veterans involving those little "buggers."

Vern Schmidt

Vern Schmidt, of Fresno, Calif., is a veteran of the 90th Infantry Division.

On March 12, 1945, I had my first shower in six weeks, and I didn't get another one until the third week in April.

That one on March the 12th was just a 10-minute shower in cold water. We were on a hillside and they said we're all gonna get showers and clean clothes, and I said, "Man, that sounds fantastic." When they took our company, my platoon – just us 12 men – we entered this tent, it was on the side of a hill, we walked in there and they said, "Take your clothes off."

Well, from March the 12th back to January I had never had my clothes off, and the food we ate, we had mostly K rations but once in a while you would butcher something along the way that loosened up the bowels, anyway, I had a good case of diarrhea for quite some time, and when they said "Take your clothes off," I thought, "Uh-oh, I'm gonna be embarrassed."

I'd found a German folding knife, it had one blade in it, when you opened it up it looked almost like a butcher knife. We took all our clothes off and I started to take my underwear off, and it wouldn't come off. So I had to use that knife and literally cut my underwear off my body.

After the shower, we got clean underwear and that was all. We still had to wear our other stuff. And I got clean socks. That was on the 12th of March. Then about the middle or latter part of April we were held up in a town overnight, and I had acquired lice by that time from sleeping in hay. We'd slept in so many places where the Germans had just left, in a lot of barns and in foxholes that had straw in in them, so we acquired a lot of body lice. They don't bother you until you get real quiet and warm, and because we wore long woolen underwear that had a lot of thick seams on it, you'd lay there real still and try to be warm and you'd feel one start over on the left side and go clear across that seam of your underwear, clear across the back and over to the other side, and you'd start scratching, and pretty soon one would start around the bottom part of your underwear and he'd go around your belly or around the back, just enough to get you scratching in the other direction.

A number of us, we were pretty lousy at that time, and we were held up in this town just long enough to get some fresh food, and one of the guys came in and said, "Hey, we're gonna have a delousing deal here, we're gonna bring some tablets and delouse clothes."

Anybody that felt they had lice could come on down. So we went down to a barn. It was night time, they closed the doors and put a candle there and lit it, there was no electricity.

There were two tubs of water and there was a sergeant, there was a corporal, and I was a Pfc. So you know who got in the water first. The sergeant got in. And I watched him, he washed his body with soap and then he stepped in the next tub and rinsed off, and meanwhile, his clothes went into a big garment bag and they stuck a bomb in there that was supposed to delouse your clothes. Then the corporal, he took all his clothes off and he stuck them in another sack and he put the bomb in there, and I was still waiting, kind of anticipating. The corporal crawled in that same tub of water and he washed off, he stepped into the next tub and rinsed off. And I thought, at home, when we were raised, on Saturday night you took a bath in a big tub in the kitchen, we didn't have running water at that time. So I thought, "Holy smokes, they don't change the water here, my mom used to do that, she'd change the water for you."

I got up there and the guy says, "Take your clothes off, Schmidt." So I undressed. I hesitated, I was the third guy to crawl in that tub of water. By that time it wasn't even hot anymore. I washed off, and I kept thinking, there's two guys been here scrubbing in there and I'm getting that same water. I got in the other tub and rinsed off. That was kind of humiliating, but anyway, when you're 19 I guess you don't care too much, or you didn't have time to think.

Half an hour later they said, "Your clothes are okay now." We got nothing clean at that time, we put that same underwear back on again, the same socks, but supposedly our clothes were deloused.

I had been given a care package that had a wool scarf, I'd wrap it around my neck at night time to keep warm, or pull it over that little cap we wore, I'd wrap it around my ears, and the guy told me to throw that away and I said, "No, that's kept me warm," and he said, "There's a lot of lice in there," but he said, "I think we got 'em killed now," so I kept the scarf.

It wasn't long before I had lice again.

John Sweren

John Sweren of Mesa, Arizona, was a tail gunner on a B-26. His plane was shot down on his 58th mission, and he became a prisoner of war. His full interview can be found in "Merry Christmas in July."

The night before I was liberated, we were in this barn, and there was a chicken place right next to it. I saw two eggs. I learned this from my father, I had this piece of wire from the bale, and I had to take some of the straw out of the baler to puncture these eggs, and I sucked the egg out, and I put the eggs back so nobody would know they were stolen. The next day, everything seemed to change. The guards were different. The same guards, but their attitudes were completely different, and we marched into this town called Bitterfeld, Germany, and here's an American tank sitting there, with some officers standing there talking to the Germans. Then people were pinching each other to see if it was real, the prisoners were.

By God it was real. We were not asleep and just dreaming. There was a feeling that I can't explain, I guess like getting married, or a miracle.

I weighed 89 or 90 pounds. After the transition took place, the German guards just disappeared, and here come American trucks. And, you know, everybody says, "What's the best thing that ever happened in your life?" Well, when I got liberated, or when my daughter got married, or when I got married or my kids graduated school. ... There's a lot of good things that happened to me in my lifetime, but one of the best things is when I got deloused. I took my clothes off, and they had kind of a big vat like, and for a long time after that I still scratched, force of habit. But those varmints, I'll tell you, there's no way to get rid of them.

Then, after that, they had some doughnuts and other things, and this lady said, "Who's in charge here?"

I said, "Nobody's in charge, really."

"Well, how about you?"

I said, "Sure. What do you want me to do?"

She said, "Get a couple of your other friends, POWs, to help you. We're gonna march people through the line here and we don't want them to take more than one doughnut. Not that we don't want to give them all to them, but we don't want them to die" – get deathly ill is what she said, not die. So we did that. One guy got through and they took him to the hospital and I don't know what happened to him. He got through somehow and ate four or five doughnuts. But anyhow, there was a Red Cross lady and she said, "What is your name?"

And I still had my big German dogtag, I said "John."

"Do you know your home phone number in the U.S.?"

I said "Yes. Why do you ask?"

She said, "It's Mother's Day in the U.S. Would you like to make a call?"

I said, "I don't have any money."

"Oh, we'll take care of it."

She dialed it for me, and my mother answered the phone. And I said, "Happy Mother's Day. This is from your liberated son John."

She said, "Who's calling? My son was killed in the war."

I finally said something that triggered it was really me, and she dropped the phone. My father got on the phone and he asked me if I had my eyesight and my limbs. I said "Yes, I was wounded some, but I'm okay. I'm a little skinny."

"What did you miss?"

I said, "Well, I missed you and mother, I missed my sisters and brother."

"Anything else?"

"I missed Christmas."

I came home on the first day of July. It was my younger sister's birthday, and there was a Christmas tree with gifts. I thought I'd get over this, feeling sorry for myself. But anyhow, I know others suffered as much as I did, the death march on Bataan, our march, I had a good friend that was on the death march in Bataan, and we compared notes.

Do I remember what I told my mother so she would know it was me? Yes. There was a food that she made one time, they call them pierogis, and I said, "I'm just anxious to have some pierogis when I get home." And she dropped the phone.

Eugene Sand, left, and Ed "Smoky" Stuever

The Jet

Eugene Sand

You usually don't know, as long as you were missed by fire you don't know how far you were missed. One of the guys on the two-ton wrecker was named Doc Bellows, and he always carried a Thompson submachine gun, and he didn't let a speck of dust get on it. We were parked for the night on an old destroyed German air base. There were some hangars along the end that were all destroyed by our fighter planes, and the runway was out in front. We parked close along the front of one of the hangars, and Doc was cleaning his Thompson submachine gun. I said, "Well, Doc, you've got your cleaning equipment out, I'll just clean a little of the dust out of my old gun. " And we were standing there, just like you and I are, that close together, we had the stuff laying on the fender of one of those big wreckers and were cleaning our guns, and all at once we hear this roar. And here comes the first jet airplane I had ever seen in my life.

We had heard of them, but we never saw them before. Our old fighter planes probably flew about 350 miles an hour, and this thing was coming in there probably 700. And smoking like a coal locomotive. And we were so amazed at the speed of that thing, we just turned and watched it. It went past, made a full circle, and came back and fired.

Dumb as we were, we stood there. And when that plane fired, one shell went between us and hit the side of the wrecker and fell down right by the fender, between the fender and the hood.

I picked it up and it was still hot, and I turned just a little bit, and I said, “My name’s not on it, Doc, it must be yours,” so I handed it to him.

We should have went for cover when we saw the plane coming, but we were so fascinated with seeing that first jet plane we had ever seen, and the speed that thing was going, we wanted to watch it. And it only took him a split second to make that circle and come in and fire. As a matter of fact, the bullets came in before the sound did.

But that one, we knew how far it missed us.

One time, early in the war, it was the evening, and we had pulled up along a row of trees, under the trees a little bit. There was a potato patch there, and the bushes of the potatoes were probably a couple feet tall. And just about that time, probably five or six German fighter planes came over, and then the American planes showed up, and they had a dogfight up there. I had seen these dogfights in the movies, and there were about three of us guys sitting and saying, “Boy, this one’s gonna be for real,” and “We’re gonna watch.” We were in this potato field, and they were circling up there and firing on one another, and diving, and all at once there was four rows of bullets went whish-whish-whish-whish through those potato vines. I don’t know the exact distance between the guns but there’s usually four guns on a fighter plane and all four fire at once, and they went on each side of us, and they didn’t hit any of us three, but boy, we never stood out in a potato patch watching any dogfights after that.

Dick Bengoechea and wife

The Purple Heart

Dick Bengoechea

Dick Bengoechea, of Boise, Idaho, joined the 712th Tank Battalion as a replacement. He was an officer of the Purple Heart Association in Boise when I interviewed him in 1995.

You can't judge a man till you've walked in his shoes. I mean, you don't know what, you do a lot of funny things when you're under pressure like that, you don't realize till you look back afterwards, especially after Malmedy.

We have a guy who was in the 82nd Airborne who jumped in Normandy and Holland and in our Purple Heart magazine, it shows us about about a platoon of German prisoners, in the snow, and there's one guy facing and one with his back, he was a lieutenant, and there's a jeep over here. And he'd just handed this guy a Thompson submachine gun and our friend that's in the Purple Heart, he'd been wounded twice, he handed him one. He says, "If one of them Krauts breaks and runs, shoot every sonofabitch." And he wrote that in the Purple Heart newsletter, and showed that picture. They had him come clear to Albuquerque just last month to get the award for the best story of the year.

Then we have another guy who was a P-51 pilot. He was low, strafing, during the Bulge. They shot him down, and he got a Purple Heart. He was on a prisoner of war train, and his own goddamn squadron shot the prisoner of war train and he got another Purple Heart. He's about 87 years old. But you find so many guys who didn't talk about the war. My kids used to say, "Why don't you talk about it?"

I said, "Well, you don't go around telling people you shot people. What is there to tell?" I found out through working with the vets, through the Purple Heart organization, the older guys, in the last year or two, have started talking about it. They get with a guy, "Oh, I was there," and "Do you remember this and that?" But a percentage of people don't understand just what it is. Like I say, until you walk in their shoes you can't ... that's what makes it interesting, I guess. We're a piece of history that shouldn't be forgotten. During World War II the whole world was involved, even the Germans. They had no economy, Hitler gave them something, Roosevelt gave us something, and everybody was in the war, on the home front, and it hasn't been that way since.

I finally talked to my son about the war about a year ago. See, I went to the VA right after the war, and they just laughed at me. I said, "I'm having awful bad problems. I can't hear." And I said, I'm not coming back to this goddamn place. Then six years ago, a couple of guys I know said, "Why don't you go down and at least get your card?" So I went down, they gave me a physical, looked my records up, and said, "Where in the hell have you been?" I'm 40 percent disabled. Then last year, on account of my hearing, they brought it up to 50. So now I get everything but teeth. All my medication, and everything. And I thought that was kind of ironic. Then they started the Purple Heart organization and I got involved with that. Then I was involved with restoring vehicles. And I've been real active since then, trying to help vets.

Another guy, who's 84, he was in the service in '38 and '39, a B-17 pilot with 27 missions over Germany. His last mission he had two engines out on his fortress, the Cliffs of Dover are 700 feet high and he's flying at 300, and the only place he can land is a little white piece of sand on the beach. So he put her in, killed all of them but himself and his co-pilot. And all these years, it's never really bothered him but then he retired, and he was really having flashbacks. He said, "I killed nine of my boys." He's got more time to think, see. You get older, and it comes back to him, remarkable, just as clear as can be. But he was an old man then. A B-17 pilot. Twenty-seven is a lot of missions.

I had another friend, this guy was a tail gunner, he went over Schweinfurt, when they shot so many of them down. And it's funny, that rear turret, the tail gunner's turret, they had to enlarge it because it was so cramped in there, and they extended it five inches, and they called it the Cheyenne turret, the reason being it was manufactured in Cheyenne, Wyoming. They made all those modifications. Well, this guy refused to go to Schweinfurt the second time, and they never did anything to him, a whole bunch of them, they just shot so many of them up. And he's got kind of a funny story he tells. He was at 10 Downing Street taking in the pubs, and they cleared out of this pub about half-drunk, his tie was off, his hat was off, and his blouse was unbuttoned, and Admiral King – this was a real sonofabitch – and Eisenhower come driving by in Eisenhower's car and King pulled it over to the curb and said, "Soldier, you're going to the MPs." Jumped out and grabbed him, pushed him up against the wall. And he said, "I'm gonna phone the MPs, you're out of uniform." Just lit into him. And Eisenhower stepped right in front and said, "What outfit are you in, son?"

He told him.

Eisenhower said, "You went over Schweinfurt, didn't you?"

He said, "Yes."

Eisenhower said, "Put your tie and your cap on and get the hell out of here." And so Eisenhower was his idol then.

Murray and Ethel Levine

"We've Got Them Surrounded"

Murray Levine

Murray Levine was a former prisoner of war. He was captured at Anzio, while serving with the 45th Division.

I'll be 76 on May 31 (1998). I was 19 when I got drafted.

After I was inducted, they sent me to Fort Dix. I was there for a week or so. They put us on a train with all the other guys, and I got sent to Camp Croft, South Carolina. That's where I took my basic training. Sixteen weeks.

When I finished my basic training in South Carolina, they took me out, because I had taken an IQ test, and I had a pretty high mark. They took four or five of us out of this training company. The Army had a program, the ASTP, Army Specialized Training Program. They were going to send us to Auburn University to do some studies or training. So we laid around for three or four weeks. Then suddenly our orders changed, and we were told to report to the training company that was in training in Croft. The Army was holding back on this ASTP, because they needed replacements in Europe. So I took another three or four weeks of basic training, and I went overseas. I had a delay en route, they sent me home, and from there I had to report to Newport News, Virginia. That's where I was shipped out. The General Anderson Naval transport was there, and I went to Africa. They shipped us to the port of Oran. We landed, and then we got off the boat, and walked across the pier to a British boat, which was actually an Indian boat. And we were going to Italy.

I got seasick on the trip to Italy because the stench of the food that they were cooking, we couldn't take it. The ship was loaded with our rations, but that they didn't give us. They were making some kind of shit, I don't know what the hell it was, a big pot, and the guy was sweating in it. And we got seasick because we didn't eat, and we smelled the stuff.

Finally we got to Naples, which was a big replacement center. I was there for about a week, and I was assigned to report to the 45th Division. So they put us on trucks, and they drove us up to an area, Anzio Nettuno it's called. And we got off the trucks and they broke us down into units. I was in the 45th Division, 157th Infantry Regiment. I was in G Company.

G Company fell back and was in a rest area, which was about a mile and a half from the front lines, and the Germans were firing artillery.

I was a B.A.R. man. Browning Automatic Rifle. Two-man team. And we were in this area, and we were talking to the lieutenant, he came and introduced us. Most of the guys in the company were replacements. You don't last long in the infantry.

I jokingly went over to the lieutenant and I said, "You know, my eyes aren't too good."

He said, "Don't worry about it. We'll put you up front, and you'll see everything." I thought that was funny as hell. We all laughed. "We'll put you up front," he says.

"Oh, good, I was worried."

A few of the guys were talking to the lieutenant, and one said, "What's the situation in the area? "

He said, "We've got them surrounded." He was full of shit. It was the other way around. The Germans had us, because they were in the hills and we were on the beaches, so how can we have them surrounded? So we were laughing. I said, "Gee, that's a funny way of surrounding."

We rested that day, and about 8 or 9 o'clock we get a call, and we have to replace a British outfit on the front lines.

So we got our stuff together, and it was 2 or 3 in the morning when we started to move up. The lieutenant said, "Don't make any noise, because the Germans will shell the roads."

As it is they were shelling, and would you believe, they start shelling the roads anyway. I was so punchy I actually walked into a tree. Banged the tree with my helmet. What a noise it made! I was half-asleep. The guys said, "How the heck could you walk into a tree?"

It took us a couple of hours, but we got to the front lines. The British outfit was pretty well decimated. Their bodies were lying all over the place. Germans. British. Americans, laying all over.

We got into the foxhole there, and it was muddy. We had mud up way beyond our ankles, and we set up our gun.

The Germans fired artillery for two days, to soften us up. On February the 23rd they lifted the artillery. I look out of my foxhole and I see them coming. Attacking. I opened up on them. Everybody, the heavy equipment, everybody who was alive was firing at them. And the B.A.R. is like a machine gun; it does a lot of damage. So what they did was they apparently surrounded my foxhole, and one of them tossed a concussion grenade that landed near it and exploded. It knocked us cold practically, and also caught me [underneath the left shoulder]. I had a wound in the shoulder. Plus a concussion. And I was just shaking my head. Then all of a sudden I heard a machine gun spraying, and my sidekick was killed and I caught a couple [of machine gun bullets] in the back. They must have gone through him into me.

The wound in the shoulder was shrapnel, that tears, and I was bleeding badly, and I was spitting blood from the concussion. I was a bloody mess.

Then I looked up and I saw the Germans were on top of me. I said to one of them, "Ich bin verwounded." And this is a crack outfit, the Hermann Goering Panzergrenadiers. Their faces were all like frostbitten. They were pinkish and bluish, you know, when you're frostbitten, you get that certain look. They had probably been on the Russian front.

One of them looks at me. He says [to his officer], "Amerikaner ser schlech verwounded."

So [the officer] says, "Macht nichts! Macht nichts!" In other words, "It doesn't matter. Let's finish him off. We're not taking any prisoners."

I hear him cocking his machine pistol ready to finish me off. That's when the thoughts were in my mind, "Goodbye Charlie." At that moment another shell exploded right among us, and just blew them away. Killed them all. Another GI was there with his hands up, out of my platoon. He was killed also. And part of my hand was blown off. If I wasn't in the foxhole, I would have been killed.

When I finally got my senses back I looked around and I saw everybody's dead. I crawled out of the foxhole. I needed a helmet, and I took a helmet off this guy, put it on my head, and I was roaming in no man's land. I was in the German lines.

There were a lot of caves in the area, and there were a lot of machine gun nests. So apparently in one of the caves [there was one machine gun nest that had been knocked out], and I said, "Kamerad." I put up my hands. A German runs out, pulls me in, and I'm looking around, and I look out and I see there are two GIs coming up with grenades. And our grenades are fragmentation grenades, if they explode in a cave, they make mincemeat out of everybody. So I started to be a pain in the ass. I started to bug the German lieutenant that I'm dying, I'm bleeding, I need help, this and that. He pushed me aside. He spoke English. He said he's busy. And I kept after him. And then he says finally, "Get him out of here." So one of the Germans took me and we went in the back somewhere where we could get out of the cave, and we were walking along, and I heard, "Varrooom! Varrooom!" The GIs knocked out the cave. Incidentally, while I was in there, it was very dark, and there were some GIs who originally were machine gunners. They were in bad shape. They were crying, in agony, they were badly wounded. But fortunately I got out of there, and they got blown away.

We got to a knoll on a hill. We're walking along, there was a big supposed first aid station, a German first aid station made out of stone, a house. And we finally got there, after a couple hours. Oh, he also took a dirty handkerchief out of his pocket and wrapped my hand up.

So I went to this supposed first aid station, and they had a lot of German wounded and Italian wounded. First I notice that they'd cut the floor out, and I see that they use it also as an observation post for artillery. And they were talking, phoning. I said, If we get wind of it, our destroyers and cruisers off the coast get will use it as a target. Don't you think they did? Fortunately the stones were [about a foot thick], so the shells would bounce off. In fact, there were two German paratroopers making a run for the house. A shell exploded and blew their legs off. One German comes over to me with a gun, he was gonna shoot me in the head he was so burned up. Another one said, "Hey, schmuck, what are you doing? Look at his condition. He's a [human] even though he's an enemy soldier." So he put his gun back.

The following morning an ambulance pulls up to take all the wounded out, the German wounded. And they're leaving me there with another GI. He wasn't wounded. So I said, "I need help."

The ambulance driver says, "There's no room."

I said, "I'll sit up front with you."

In fact, the other guy said, "We'll both sit up front with you."

"Nein." He says. "You got cigarettes?"

I said, "I don't know."

So he frisked me. Fortunately I had cigarettes. I never smoked, but from the 10-in-1 ration I used to get them. He said, "You've got Chesterfields. Don't you have any Lucky Schtrik?"

"I'm sorry. Just Chesterfield." It shows you the tobacco, how important it was.

He took me, and the other guy. We sat up front. We squeezed in somehow.

There was another German medic, he stayed on the running board. And we were riding along in the ambulance and he's going through a tank area. Don't you think we get attacked by American planes? The driver stops and parks his ambulance near a tank. The GIs are going for the tanks, they're shooting up the place. They accidentally hit the ambulance too.

Before that, the German medic who was standing on the running board jumped off, and he got hit by an anti-personnel bomb. I saw him take off and he never came down. He just disappeared.

Then, after the attack, the driver was looking around. He saw the ambulance had been hit. He looked in the back. Some of them were killed. But fortunately me and the other guy were all right.

So the driver took us to Florence, Italy. He took me to a hospital. I don't know what happened to the other guy. And I was there in that hospital for three or four days. They gave me very little medical attention. In fact, they wrapped my hand up with paper bandages. Can you imagine? And they put me on a hospital train, and during the night they dropped me off at a prisoner of war camp, which was Stalag 17.

"Russky Gonna Cut You throat"

Ed "Smoky" Stuever

In Normandy, they liberated a young Russian. He was about 21 years old, and the Germans had him digging foxholes. So when he heard the Americans were coming he hid out, and he waited till we came along. They told me to take him with me, he's a good worker.

I had him with me doing hard work, moving the tracks around. We had him from Normandy until August, when we were in Briey, France. He was walking down the main street over that little bridge, and he saluted a colonel of the MPs in a jeep with a cigarette in his hand, and they hauled him away. We never saw him again.

In all this period, he would always approach me and he'd say, "Sergeant, what you do when Russia and America come together?"

I said, "Oh, I go home. I can go home tomorrow, I have enough points."

"No, no. You no go home. Russky gonna cut you throat." Time and time again he would approach me with this, "Smoky, what you do when America and Russia come together?"

I couldn't sleep with that guy around. I had a real sharp dagger, and I had an extra pair of boots on the truck. They were all worn and I wanted to exchange them whenever I got a chance, for myself, and he said, "No, I want them."

I said, "No, you can't have them."

He said, "Let me have your knife." This was still in Normandy. And the next day he comes back with a shiny pair of Nazi boots on, and he gave me my bloody knife back. So I could never sleep with him around.

"Are You a Jew?"

Murray Levine

They brought me at two in the morning to Stalag 17. They dropped me off on the Russian compound side. The lights went on and Russians dragged me in. They treated the Russians like, I can't believe it. And I'm looking around. They took me and dragged me in a bed which was too small for me but I lay there. And I'm looking around. I see Russkies. I don't see any GIs. And I hear a lot of Russkies talking. They're in bad shape, all of them. Their heads were shaved. The Germans treated them like dirt.

Finally, everybody left the room except one guy. Did you ever get a feeling that somebody's staring at you? I look around and I look at him, and he smiles at me. I smile back. He hobbles over to me. I see that he was wounded in the leg; he had his kneecap shot off.

The first thing, he looks at me, "Du bist Amerikaner?"

I said, "Yeah."

He said, "I'm Russky."

I said, "That's what I figured."

He says, "I wish I was an American." He says, "Du bist ein offizier?" Am I an officer? He spoke in German.

I said, "No, I'm a private."

He says that he's an offizier. "Leutnant." Russian Leutnant.

He looks at me. He says, "Du bist ein Jud?"

I said, "Yeah." I'm a Jew.

He says, "So am I."

He says, "The next time I go to the lazarete" – that's the delousing chamber – "and they see that I'm circumsized, they're going to kill me."

For a couple of days, I was just heartbroken. Finally, while he was talking to me, a couple of GIs come in. They're looking for me, because they had a roster, and they expected some GIs to come in, and my name was on there. They figured somebody screwed up, they must have dropped me off in the Russian compound, and they found me. They said, “What are you doing here?”

I said, “I don’t know.” How the heck should I know?

So they got me and they took me to the American compound. In the interim, I befriended this Russky, and I used to make sure that I would give him some food, because we used to get a Red Cross parcel, which was actually an American Army parcel but they were using the Red Cross as a front.

His name was Liebman, I’ll never forget.

Then one day we were looking for him, because I couldn’t find him. The Germans had killed him. Only because he was a Jew.

Klaus Schilling

The Death Camp Doctor

Darrell Petty

We liberated the Flossenburg concentration camp. And the 4th Armored and part of our unit went to Buchenwald, too. But you won't find it in any history book, because we were stopped at Merkers, where the gold reserves were. And Patton said it doesn't take us all to watch Merkers, and there was a little place called Ohrdruf. It was supposed to be a work camp. But we got there and there were bodies stacked up. And then we went to Buchenwald. The Germans were gone and the prisoners had apparently taken it over by that time, but we saw what was there.

After the war was over, I stayed in the army of occupation and they transferred me back to Munich. I was attached to the 508th MP Battalion. And we pulled regular MP duty. But we also were at Dachau, at the war crime trials, and at Nuremberg.

One of the defendants whose name stands out in my mind was old Dr. Schilling, because he did so much experimenting on the people. He was the camp doctor at Dachau. When we took those places, it was horrible to see what was there, but we still didn't know about all the experimentation until the trials. And I came within an inch of shooting him.

We had some MPs that had never seen an ounce of combat, who came over after the war, and we had one new guy there, I won't mention any names, but he'd never seen combat. He had a little .30 carbine, they were semiautomatics at that time and had 15-round clips, but there would be a clip in the gun, and two clips on the butt of it. That's 45 rounds. He's standing there gawking around and he wouldn't be protecting that gun. I got on him several times. I told him, "You protect that firearm. These guys don't have anything to lose."

Well, these guys that lived such a high muckety-muck life, it was kind of satisfying to see them sniping cigarettes off the floor, and they were eating some pretty thin soup that we brought for lunch, and they were in a soup line. And this kid was standing there and he's got the butt of the rifle on the floor, and he's not looking at them at all. I just about went over and said something to him but I didn't. And by golly, old Schilling was in line, and I, I just loathed him for what he'd done to people. And all at once he made a dive toward this kid, and the first thing flashed in my mind was he's going for that carbine. I had a .45, and I always had it fully loaded, the hammer on half-cock and the safety on.

And he dove like he was reaching for that rifle. As it turned out there was a cigarette butt about half-smoked, boy, that was a prize, between that kid's foot and the butt of the rifle. That's what he was going for. But I didn't know that. And I grabbed the .45 and I dropped the safety and cracked the hammer full, and when he came up with that cigarette, I had the .45 about six inches from his face.

And his color just drained, "Oh, nein! Nein! Bitte! Bitte! Zigaretten, Rauchen! Rauchen! Bitte, nicht schiessen!" Don't shoot. Please. Rauchen is smoke. And I never had a feeling like that in my life, before or since. I wanted to pull the trigger. I couldn't hardly keep from pulling the trigger. And I finally just pushed my finger off of it. And I dropped the safety on, and I grabbed him and boy I threw him back in the line and I told him to stay there, and I did a pirouette and I kicked that kid just as hard as I could kick him right in the behind, and he lost his grip on the rifle and I grabbed it before it hit the floor, and he went down. And I stood above him, I had that butt of that rifle right in his face. And I told him in no uncertain terms what I'd do to him if I ever saw him be careless with the gun again, and I said, "Much as I wanted to do it, you almost made me kill a man without a reason."

Darrell Petty was 19 years old at the time this incident occurred. Dr. Klaus Schilling was found guilty of all charges, and was hanged in 1946.

The American flag that Len Goodgal placed on the cliff at Pointe du Hoc to help the battleships in their targeting on D-Day.

A Walk in the Sun

Len Goodgal

Paratrooper Len Goodgal of the 101st Airborne Division, landed in the water in the early hours of D-Day, and wound up scaling the cliffs of Pointe du Hoc with the Rangers.

The first Germans I saw were out in Breckenridge, Kentucky. We were on maneuvers, and they had prisoners of war there. We were out in a field sleeping in pup tents and these guys were in barracks. That didn't sit well with us. They had Italian prisoners out there, too. I guess they were from North Africa.

We were being prepped as superguys. Guys who could beat anybody. You can do anything. They were psyching guys up. A lot of kids thought they could; they'd go out and get into fights, MPs would pick them up, and they'd let them get away with it.

At Fort Benning, the tankers had boots, and paratroopers wear boots, and they developed a rivalry. In the mind of an 18, 19, 20-year-old kid, this is big, important stuff, those guys can't wear our boots!

I was always smart enough to know that when a guy's bigger than you, you don't mess with him, I didn't care whether he was in the Marines or the artillery, infantry, or whether he was in a medical battalion, if a guy was bigger than you he was stronger than you. When you get to thinking that you could overcome all that, and you take on anybody, you're in trouble with yourself. We had kids who tried it. Guys would come back all beat up, "Oh, what a great time I had!"

I learned to avoid that. I avoided it mostly because big guys could beat little guys, all things being equal. You usually found two guys the same size going at it. We had a couple of buck sergeants that were pretty tough. They used to invite guys out in back of the barracks once in a while. I saw one of them invite a guy out in back of the barracks and the guy walked back in front, he left the sergeant laying back there. He got the wrong guy. But, you know, these guys who came into the service were civilians, and they were old Army guys, they could beat anybody. Sometimes you pay the price for that.

To a lot of soldiers, being in the service and fighting a war was a walk in the sun. Everybody in their life needs a walk in the sun, something nice to happen to them, to carry them through life. Everybody does. Whether it's something your mother did for you, your father did for you, somebody did for you, some woman you love, something in your life made you feel good about yourself. Everybody should. That's what's wrong with a lot of these kids on the street that want recognition. They're not having a walk in the sun. They're having a walk in the night. It's a nightmare. And it shouldn't be. Everyone, no matter who he is, whatever he looks like, what color or religion, should have a walk in the sun, and that's a wonderful thing. It's one of the reasons why when you have a marriage they have a big wedding. If it's a poor family, to have a big wedding, that might be the only beautiful day that woman will have in her life. That's a walk in the sun too, isn't it? And they can always look back at those pictures and look how beautiful they are.

For a lot of these fellows that came out of the South or the Midwest or even the cities, poor homes, they went into the service and they had clothes and food, they felt good about themselves. That's why they joined up. It's like the parachute infantry, or the tanks, or the Marines. They wanted to feel good about themselves. They wanted their walk in the sun, they didn't maybe realize it, we didn't realize it, but it was a walk in the sun. It was our great day.

I came from an old Jewish family. I had a sister and two brothers. My sister was married. My father wasn't rich; he was a tailor. We all tried to be somebody. My brother was going to college. One brother was going to the University of Maryland. My other brother was a Naval Air Corps pilot. The other one was in the Army, too. I don't know whether he was in an engineering battalion or the infantry. In fact I met him overseas in Berchtesgaden when the war was over. He came to visit me just before I went home. He went home, he got married when he got home. He just retired as a professor of microbiology at the University of Pennsylvania. When we were kids he used to hitchhike 32 miles to school and back, to save the carfare. He became the head of the microbiology department at the University of Pennsylvania Medical School.

Everybody should have an opportunity to make it. If somehow you could just bring it to everybody, a walk in the sun for everybody. That's what I believe in. It doesn't have to be material wealth. It just has to be that image, that you are something and somebody and you feel good about yourself, you can always refer to it.

I volunteered for the paratroops. When I went into the service, kids were volunteering for the Navy and they were volunteering for the Marines. I was looking for something that would make me feel good about myself, and the airborne sounded good. I'll be better than the next guy, I'll do something more daring. It sounded interesting.

Actually, you're an infantry soldier. You didn't think of yourself as one in those days. You were a paratrooper. But you're an infantry soldier. It's just a way of getting there. If they shot you out of a cannon and put you there you'd have done that, too. It's just a way of getting there. It's all transportation.

Our training was tough, because they were trying to weed out people, and they wanted you to have endurance and be able to take abuse. They had a lot of ideas about what really constituted a brave guy or a solid guy. It's all an individual thing. Some of the biggest guys do some of the bravest things and some of the littlest guys do some of the bravest things. It really fell upon the individual. Their estimates of who was gonna be brave and who wasn't, if somebody got a lot of blisters and fell out of a march, they'd kick him out of the outfit because he didn't want to kill his feet with the blisters. But at the same time he makes a more sensible guy than a guy that keeps marching and kills his feet. That's the way it goes.

I remember guys saying, "Who the hell needs this? Kick me outta here. Kick me the hell out." Some guys just haven't got it. Most of them were psyched up for it.

The first time I jumped was easy. Of course you'd been prepared for it. We made mockup jumps. Some of the mockup jumps were harder to make than the regular jumps. You fell down, you're 25 feet from the ground. That's a little harder to do than looking out of a plane, there's no ground, there's no feeling of height, and out you go. Once you do, and you get the feel of it, when you come down and land it's a grand feeling. You feel like you've conquered the world. You've really got the world by the balls, that's the way you feel. But the truth of the matter is that it's just an elation. You're so elated. I saw very few guys get hurt. Some guys did. I never saw a chute fail.

I heard a story about a guy in Normandy who had his static harness cut by a piece of shrapnel, and the chute never opened and they found him on the ground like that. I don't know if it's true. But it's a story that I heard. When you hear something secondhand, you don't know if it's true or not, because some of them have vivid, vivid imaginations. As the years go by, you will tell yourself things about yourself, and you begin to believe them.

In Normandy the casualties ran heavy for a week. Once we got stabilized, we were in pretty good shape. In Holland, they were running heavy all the time. Actually the first day in Holland was our easiest day almost, because they didn't defend the jump field that we were in, we just took it over. There was nobody there. We just jumped down and took it over, and we set our security for the division. They blew the bridge over the Wilhelmina Canal. We went over on ropes at night. Hand over hand on ropes. That's the way the whole battalion went over and they took Eindhoven the next day. It rained that night. God damned drizzly. Real bitter at night.

On D-Day I got wet. They hit my plane as we went over the coast. I remember going over, I was looking out. I could see the sea. I could see the coast from where I was in the plane. I was looking out the door. The door was open. It was a beautiful night. Twelve o'clock, 12:30, something like that. We went over the coast, all of a sudden, Christ, there was anti-aircraft fire, and fire, we could see it on the ground, in the air, in the sky. And the plane got hit in the tail. At least we're ready to jump. The plane got hit in the tail and went into a dive. I just stood there hanging onto my static line thinking I was gonna get killed. I didn't really think of anything but hanging in there. And it pulled out of the dive. When he pulled out it went up a little bit, and the lights – when it went down it turned the lights red instead of green, or yellow. Then they turned it back to yellow. I went around and got back in the door. "Everybody get ready to go!" Boom! All of a sudden we left.

We got hit again, in the right side. I didn't know where. The fellow that was No. 2, Nils Christensen, told me that the right wing was on fire. He could see out there better than I could, that the right engine was on fire. We must have gotten hit in the right engine, and it was tilted that way. We were coming out the other way, so I just about got out of the plane. I was the fourth man, and the last guy out.

I had a lot of equipment on.

The plane was going down. I knew it was going down. I didn't know where it was. I knew the plane was hit and it was sinking, not too rapidly either. Of the four guys who got out of the plane, two of them landed on top of the cliff, and two of us landed in the water. Me and another guy landed in the water, and two landed on the cliffs. One of them was Nils Christensen, who was captured. Another one was Lieutenant Johnson, who got hurt, and got back to the outfit somehow or other and was evacuated. But Nils Christensen spent ten months in a prisoner of war camp.

I hit the water. I didn't know whether I was going to drown or what. I landed right on my back almost. I took my chute and snapped it off, got out of my stuff and got my rifle, which I had strapped across me. My equipment was underneath the chute when I got up.

I started to swim, but when I stood up I was only in about a foot of water. The other guy was calling me. In fact, he was saying, "Is that you, Sam?"

I said, "Yeah!" My middle name is Samuel, all the guys called me Sam.

He said, "Where are we at?"

I said, "I wish I knew. I hope this is the White Cliffs of Dover," because I saw those goddamn cliffs. How are we going to get up these sonofabitches? We went under the cliffs looking for where the hell we were. We didn't have any maps, nobody told us about those cliffs or anything else.

So we worked our way along the shore. We couldn't find a way up. We tried to get up one place, it looked like we might be able to, but we couldn't. We got down a little further, and they were bombing. The bombers were dropping 2,000-pound bombs all night long. And the Navy was shelling the cliffs. We actually put our gas masks on, and the canisters didn't work, we couldn't breathe. Because I smelled the powder; I thought it was lewisite gas, which smells sweet. I figured maybe we're being gassed with that powder burning when it hit the cliffs. What the hell's going on? And we didn't know. We threw the gas masks away because they didn't work.

Raymond Crouch is the guy who was with me, Raymond L. Crouch from Richmond, Virginia. He was a sweetheart of a guy.

In the morning, early, we see these boats coming along the cliffs. I said, "Hey, maybe they'll come in and get us."

Well, as we walk along, all of a sudden they're coming in and assaulting these cliffs, firing the toggle ropes up and all that. These guys are running and jumping and going like crazy, and on top of the cliffs they're cutting the ropes, firing down, and throwing grenades over the side. We're down there and the colonel's motioning for us to come over. Colonel Rudder has died since then. He was a nice guy. No Ranger ever got the Congressional Medal of Honor, but he deserved it. He was recommended for it.

I didn't have a high regard for the guy until that happened. Then I realized what they had done.

The guns were not in place there. Guys tell me they were back in the woods, and that they put grenades in there to burn them out. But I recall seeing two guns along the road, two big long guns. They didn't look burned out to me. I saw two of them. Maybe they were. They were camouflaged. I thought that's what they were gonna use up there in those emplacements, the gun emplacements, but at the time we didn't know. We really had no idea what it was. I didn't know what I was going for. I didn't know what I was doing up there after I got up there.

The next day – the first day I saw a lot of ships out there, but the second day it was a sea of ships. You couldn't see water anymore. Just a sea of ships like foam churning on the water. All you could see was the battleship Texas, which was our artillery. A couple of little destroyers. And then there was a sea of ships, all white. You could not see the water past where the ships started. It was like that for a couple of days. So you know if they had those guns up there, the damage they could do. You couldn't see Utah Beach, it was too far.

You couldn't see Omaha Beach to the right, it's too far. All I saw was cliffs. What a bitch. Those guys really were heroes because they knew what they were going for. I was just incidental. They could have been giving out candy bars up there, what the hell did I know was up there.

They threw everything over the goddamn cliffs. They didn't start really getting at us until we started getting up there. As we came up they became aware, but if that thing had been heavily defended, nobody could get up there. And the other thing was, once we got up there, everybody had a B.A.R., a machine gun or something. I picked up a tommy gun. Who the hell wants a rifle? You're firing away at everything you can see. Some of these guys had to attack pillboxes to get them out.

We were up behind a pillbox. The colonel was with us. We were getting fired on. They needed artillery support, and they wanted the Navy to fire over that point. They wanted somebody to go back and get the flag that was hanging over the cliff, but who's gonna go? I said, "Okay, I'll go." So I went to get the flag, which was hanging over the cliff. I got the flag. I got fired at going out to get it and coming back. The colonel commended me highly for going and doing it.

The ships could see the flag and use it as an artillery point, and know where we were, and fire beyond it. We needed it.

The ships were our artillery. We had a Naval observer with us who came up with a radio. We used the radio to tell them where to fire.

The casualties were so heavy, all we could do was just sit in there and hold on till they broke through to us.

Guys got heads blown off. Broke a leg or arm falling off the cliff. We had prisoners underneath the cliff that they sent back by boat; they brought in a boat, I don't know whether the Navy sent it or it was a boat they brought in, to take them back. We didn't want to keep them there, the German prisoners. For three days we were just getting prisoners. In back of us. They'd pop up, you didn't know where they came from.

They'd usually surrender. They knew how to surrender. They'd unbutton their overcoats and throw their helmets and rifles away. I never caught a German soldier with his helmet on unless it was right on a foxhole or something had happened. Otherwise they knew how to surrender. They'd take their helmets off, unbutton their overcoat and throw their rifle away. They'd say "Kamerad, Kamerad."

Some of them were Germans, but there were also Polish. There was a Polish battalion of workers that were there to do that work. They were helping to build the fortifications. But they had German uniforms, green uniforms, Wehrmacht.

They didn't seem that young to me. Most of the soldiers that I saw looked like battle-hardened guys, or had been around a while. Or they were like engineers' workers, they could have been any age. They weren't that old.

I got wounded on the 13th of June at Carentan. We went back to my outfit after the fifth day, on the 11th, so I was there from the 6th to the 11th. I went back to my outfit the 11th, that night. And the 13th we made an attack on the other side of Carentan, into a German counterattack. They wanted to take the town and we held them off. That was a brutal battle called Bloody Gully. Our company, I Company, called it Bloody Gully. We lost a lot of men in there. Jim Schueler was talking about that, from California. He died a few years ago. And he said that he thought it was the most brutal battle he had ever been in. We lost a lot of men. In fact I shot a first sergeant up there, from H Company. I think it was G or H Company, a Sergeant Holmes, first sergeant. They were out of position. We were firing all through that place. The company commander sent me up to get a squad. He'd sent a machine gun squad for flanking fire up there. We hadn't heard from them. He said, "Where the hell did they go? Go up and see where those guys are. What the hell is happening to them?" We wanted some flanking fire. We're getting it coming this way and we had no flanking fire. So I went up and went across, Jesus Christ, these guys are popping up and down in a ditch, in the hedgerow. I said, "Raus, Schnell! Come out of there."

No answer. I was gonna take off, but I can't take off, they'll shoot me in the back. So I jumped in the ditch firing my rifle; actually, I had a tommy gun. Jesus Christ, our guys were digging all the way in the ditch there with their hands and all trying to dig a foxhole. I said, "What the hell's going on here?"

"Watch it! Watch it! There's a sniper up there."

I said, "Where?"

He said, "Up in the trees."

I said, "Did anybody get hurt?"

"Ohhhhh," he says. He's holding his back.

I thought, "Shit, I killed this poor bastard."

I pulled up his shirt. Did you ever burn yourself on a pot? That's just how it kind of grazed him. I couldn't believe it. I'd just grazed him, and I saw that. Was I lucky! They were in front of our position, we didn't know it. We didn't realize it was G or H Company's headquarters. They were out of position. I don't know how they got over there that far but they did. But it was in the morning, we started early in the morning. It was dark, so they might have gone this way and that way, and we were firing I guess, I can imagine those guys were getting it from us and them and everybody else and they were firing back at us. They didn't know where they were. We didn't know who they were. I didn't know. I jumped in there, and when I saw who it was I almost died because I figured I'd killed one or two of them. And the only guy I got was him. It just grazed him on the back. He pulled his shirt up and I saw that, and ohhh, almost tears came out of my eyes I was so happy.

I said, "You're okay, Sergeant, it just grazed you."

Then I said, "Anything else?" I looked all around.

I said, "You're lucky. You've got a million dollar wound."

Then I took off. They came and got the guys over about 150 yards on our right, and they moved up to position. They couldn't get up any further, they were getting fired on.

I told them, "I don't give a god damn what you guys do, don't fire in that direction because it's H Company's headquarters, or G Company's."

They said, "Christ, they've been firing at us."

Then they gave us a little flanking fire, and we stopped the counterattack. Once you stop that German counterattack it was all over. They weren't gonna take Carentan back and that's all there was to it, and they knew it, and that was the turning point. As I see it, it could have been a turning point in the campaign, Carentan was the linchpin between the two beaches. They weren't gonna take that. We were pushing out from there, and meanwhile the 4th Division was moving stuff in and they were bringing troops in. They were coming in pretty heavy. So by that time it was all over. And then we had air superiority. They just had a couple of planes, at night you'd hear them come over. And their engines were never synchronized, RrRrrrRr, they were propeller engines, just for not being synchronized.

I was wounded by a piece of shrapnel. It hit me in the knee. An 88 shell hit a tree, a treeburst, and I got a piece of shrapnel.

Being wounded is not what you think it is. It's frightening when you first know you're hit. Then you realize you're not gonna die, it's okay. But when you first get hit, it's really frightening, to see the blood squirting. You don't really hurt, you're just frightened. Then all of a sudden you realize you're not gonna die, and you're okay. Just another incident.

I crawled back with a bunch of guys, about 20 of us. In fact, a captain at the time said, "All you guys that are wounded, go back to the aid station, get fixed up." Because we didn't have any medics left up there. We had one guy up there. So we went back to the aid station, about 20 of us, and some of them stayed. I couldn't. I had this shrapnel in my knee and they had to take it out. So I went back to England and they took it out. No big deal

I came out of the hospital and went back to my outfit. I went back to the camp. They weren't back from France yet. Most of the fighting was over. They came back to England, and they began to prepare for whatever they were gonna do next. They had a half a dozen different things they were going to do when they decided to go to Holland. They had dozens of jumps planned, different things that we could have done. But hey, they picked the opportune thing at the time. Battles change. Lines change and you have to have plans for any kind of contingency. As I understand it they had numerous ideas for jumping the guys. That one took preference.

In Holland, they didn't recognize that the Germans were up there like that. They could have gotten enough antitank stuff up there to knock out some of that armor, because it's tough as hell for infantry to fight armor in open spaces, on flat land. So if they could have had some more antitank guns up there it might have been an ideal situation because they engaged a number of armored divisions, panzer divisions. If they could have engaged them and disrupted them enough they could have put them out of commission. You can't replace an armored division that fast. You've got to train a guy to run a tank and maneuver it and everything else. Once you knock them out it's different than knocking out a soldier, you get another guy to stand in his place. He's just carrying a rifle or a weapon.

I was a Pfc. I was a rifleman. But you did everything. I worked out with a mortar squad. I had machine guns at times, I had rifles at times, and I had a bazooka one time. You were prepared for any kind of weapon and at some time or other if you're in combat long enough you use almost all of them. But mostly you had a rifle and that was it. If I could get hold of a B.A.R. I loved it, but you had to carry a lot of ammunition for those things. They were heavy. They weighed 21 pounds apiece or something like that. A lot of guys didn't like hauling them around.

In addition to Normandy and Holland, I fought in Bastogne. But on the line. I went on patrol with three guys, I remember, Steele and Dicky Shinn and an officer; Lieutenant Johnson was a new officer, and they told him to take a patrol out, so he looked for some volunteers. Nobody volunteered. So I said, "I'll go with you." I don't know what the hell was in my mind. And then these other two guys said, "We'll go if you go." So we went. We went across an open field. Snow on the goddamn field, we had to go across a field to a hedgerow ditch, and we went through the woods, and we're observing the Germans with the tanks back there. Where they were and what they were doing. And then we had to come back in. You go on patrol like that, they were shooting the shit out of the place. We came up a ditch on the right side of the road. Cold, mud up to here. We were glad to see it, too. We crawled right through it. It was unbelievable. We all got back.

Dicky Shinn is out in California someplace, married and has a lot of kids. He was in the Korean War too. He was a Korean kid, that's what he was, and was a boxer. Excellent boxer. He was one of the ETO champs. He was an American boxer, well-known kid. Dicky Shinn was his name. Nice guy. I saw him in California. I saw him out in Reno. He gave me a belt buckle with the 101st insignia on it. Came with his wife. That was the first time he had been in combat and he went with me and I got him back, so he felt pretty good. I don't know what happened to the [commanding officer] after that. I was wounded. I had trench feet up there and I got wounded, too. I got hit in the face. I was below the ground and raised my head. They were firing past us. One bullet went past my face and got me right [in the cheek]. But the trench feet is what got me out of there. I couldn't walk any more. We never took our shoes off, and we were in the water, mud, freezing cold. I think it started in Holland, my feet were hurting me. I'm in the ditches up in there, foxholes filled with water, we used to dig through the dikes and set up machine gun positions through the dikes. Cold as hell and the sleet would come across the dike and cut your face practically. You'd try to keep warm. Ridiculous. It was tough up there. Hey, it was all part of the act. Be on guard duty two hours with the machine gun at night, on two hours, off, and another guy would come up. In your platoon, you would switch guys on the machine gun at night. I'd be up one night and the next night another guy would be up.

After I returned to the outfit [from suffering trenchfoot at Bastogne], we liberated Landsberg Prison. I didn't see the prison. I saw the prisoners. I've got an interesting story about that.

When we liberated Landsberg I was right outside the town; the prison is part of the town, and I was in a machine gun position with Eddie Austin. We had three of these refugees with the stripes. They were Russian I think. I don't know whether they were Jews or not, but they looked horrible. They looked nightmarish. And we were trying to feed them something. We had D-ration bars we were putting into our canteens and heating them with water, and we had some pickled eggs we used to find in the basements, and some bread that we grabbed from someplace, and a couple of things, I don't know what, we'd find stuff in cellars. These guys wanted something to eat, which we were trying to feed them, some of the cocoa. And this one guy was shaking terribly. Eddie Austin took his coat off and gave it to him. And Jesus, I almost flipped, I said, "Are you out of your mind? What are you doing? Giving away your overcoat?"

He said, "I can't not give it to him."

I said, "What do you mean you can't? They'll bust your balls for that."

He said, "Who needs it more, him or me?" He or I. Him or me. "Who needs it more, he or I?" is correct. He said, "Who needs it more, him or me?" Eddie Austin was a superintendent of schools in California later on, but that's besides the point. He knew what he was saying, him or me? He knew what he was saying, "Who needs it more, him or me?"

"I don't know, Eddie," I said. "They're gonna bust your balls."

Until my dying day I'll never forgive myself for not giving my coat to one of the other guys. But they weren't shivering. This guy was really shivering. I didn't feel bad. I said, "I think you're out of your mind, Eddie, what's it gonna do for him? Put him in a house someplace and build a fire." We had a fire going there. We fed them. They couldn't eat the stuff, it was too rich for them. The Red Cross got ahold of these guys and took care of them. In fact, the Red Cross MPs came into most of these camps. At Dachau they got in right away. They didn't let us in. We went right past Dachau. You would see guys roaming around. We picked up a guy going through Germany, and he went all the way to Berchtesgaden with us, and the cook took care of him. Maurice. I don't know what his last name is. He spoke about five languages. He'd been in a concentration camp for years. He was a skinny guy. By the time we got to Berchtesgaden he [had put on a lot of weight]. He had a horse and wagon, and anything he could steal he put in there. He and the cook were, I don't know how they were doing it, but they were getting enough loot to live for the rest of their lives. Maurice was helping him. And the cook used to carry a .45 pistol on him, all the time. And he was a pretty good guy, just gung-ho, till he became the cook.

How he got to be our cook. ... We had a guy named Feoli who was a cook and he shot his dick off in Holland because he fired his pistol while he was cleaning it. His .45 went off and it shot his dick off. I remember that, and oh, we laughed about it. What is it, pathos or something? I don't know what you call it. It's funny as hell but when you think about it, in other words it's not funny. Oh, it was brutal. We'd look at each other and laugh. "Don't give me no pistol." We used to pick up pistols, P-38s and Lugers and all different kind of weapons, Schmeissers and stuff. Most of them disappeared or we traded them along the way.

Some of the guys came back with them. Most of the guys didn't even come back with a pistol, on account they were afraid they'd use them. Some of them just didn't want to have them around. Weapons didn't mean anything anymore. We were glad it was over. But I remember us being in Solfelden, and they had horses there and we used to go horseback riding in the afternoon or in the morning. They had guys that knew about horses taking care of them. To me it was a novelty. It was recreation.

The war was over May 8th. We were going through Germany, I was there from May till September, and hey, it was a walk in the sun being a soldier. Just imagine how we felt over there. Anything you wanted, anything you wanted to do. If you wanted to fraternize – and you weren't supposed to fraternize with the Germans – everybody had a girlfriend someplace. Everybody. I can tell you an interesting story. I read Solzhenitsyn's Gulag, I don't know if you've read it. You've got to read it. It's brutal reading. And in it he talks about them returning slave laborers, women, back to Russia, and the Russians put them in concentration camps, sent them to Siberia. When I was in Berchtesgaden, we used to use them, they were working as cooks and stuff for us because, hey, the Germans don't want them anymore, they got captured, they were working in the fields at home, in the hotels, hey, so the government took care of them. Then one day they came along, they took them all, put them on trains and sent them back to Russia. Solzhenitsyn talked about them, taking those people and putting them in Siberia. And I didn't know. I thought they were going home. And they didn't really want to go home, a lot of them, because, hey, when you see Russia, what it was about, and you see what Germany was like in Bavaria, even with the work that went on. Hey, it was a better life in a lot of ways, and some of them lived with German guys.

We had thousands of troops surrender to us coming off the Russian front. They wanted us to turn around and fight the Russians. They came in with all their equipment, ready to fight. They would turn around and they were gonna fight with us; they put them in camps. We did not abuse them. They talk about starving them to death. Eisenhower did not do that. Whoever wrote that stuff is a brutal liar. We did not brutalize the Germans, although there was high unemployment among them. We did not starve or brutalize them. We weren't allowed to do that.

Landsberg Prison is where Hitler got his start, I think. They had the putsches there. Beautiful area. Berchtesgaden, and that area through there, is probably one of the most beautiful parts of the world. You have to go to see it, in July and August, when it's warm and sunny and the fields are growing, there's snow on top of the mountains and it would melt and we'd come the next day and clouds would go over, there'd be snow on the mountains. A gorgeous countryside. And they kept it nice, too. The Germans are meticulous. Clean. You can't believe that these people were Nazis, and how brutal they were. Committed all these brutal acts. When you know the people. Germany was a country of kinder, kirche, kuchen. Children, church and cooking, that's what a woman's job was. Children he made, women and children for the Reich, screw the church, and fuck the soldiers. Hooray for Germany. Blond and blue-eyed. Hitler was a brown-eyed, black-haired sonofabitch, who the hell is he kidding? It was ridiculous. It was so ridiculous that you wonder how they put over what they did. They did it. They captivated the state, captured the times. They were in a great Depression, saw no way out, were looking to blame somebody and anybody they could put their hands on, it was their fault. Never my fault whatever happens to me, it's your fault. You're to blame. Not me. If I want to believe it, I can believe it. People can always believe anything. That's what happened in Germany. The Depression. They lost contact with the reality of what kind of people they were.

An Open Heart

Dale Albee

Dale Albee, of Prospect Lake, Ore., enlisted in the horse cavalry in 1938. He was one of 14 sergeants in the 712th Tank Battalion who received battlefield commissions.

When I came back from the war, people would ask you, "Well, how was it?"

And you'd start to say and then they'd tell you about the gas rationing and the shoe rationing, and finally you learned to clam up and say, "Ah, it wasn't bad." And you just didn't talk about it anymore.

One night my father sat me down and he said, "You know, Son, we've always kept an open heart, and if you want to tell me about it, I'll listen." He said, "I'd really like to know what it was all about."

And I sat there for four hours and told my father things I'd never told anybody. I really opened my heart and I think maybe that might have kept me from going off my rocker. I was just so glad that my father had done that to me, and it shaped me up, so that I was able to do it for my daughter. My daughter went to Vietnam and spent a year over there. She was a nurse, in a M*A*S*H unit.

She came back sick, and was still scared. And the family couldn't understand, what's wrong with Donna, she's so different? And I had the chance to sit down with her and do the same thing my father did.

She went over and spent a year in Chu Lai, in the hospital. We talked, and she said, "You know, Dad, the only thing that bothers me, of all the people that came through that hospital, I don't remember a name. I can see faces, but if I go to the Wall, and I'd like to go to the Wall, the only one that I know that would be there is Sharon Lane, and she came in after I left." She's the one that came in as a replacement and got killed in that rocket attack.

Bud Painter, left, and Sam Trinca

Gold Out of a Dollar

During my conversation with the group of veterans of the 299th Combat Engineer Battalion, a sixth veteran, Jim DePalma, joined us. Jim said he had two brothers who were in the service and that one of them, a glider pilot, was killed in action. At which point Chuck Hurlbut said, "Sammy, you met your brother over there, didn't you?"

Sam Trinca

Speaking of brothers, now that you mention it, my brother was, oh, he had a job! He was in the SHAEF (Supreme Headquarters of the Allied Expeditionary Force) headquarters. He was doing the payrolls of all us guys. That's how he found out where I was. By looking at the records, he found out that I was in Nuremburg.

"I was pulling guard duty at the house we were staying at. All of a sudden a jeep pulls up and my brother gets out. He's all dressed up, and here I am all dirty, like a bum. I look at him, he looks at me, I said, "What the hell, I must be dreaming." So we took pictures that day. And he asked the sergeant, "Could you let Sam go for a couple or three days?"

"Ahhhh," the sergeant says, "I don't know."

"Jeez," my brother says, "we haven't seen each other for three years."

"Wellll, go ahead, you can go," he says. "So I didn't even bother changing. I had the clothes I had on. I jumped in the jeep and we took off, and their headquarters is Salzburg, Austria. He took me down there. A guy comes out, opens the jeep, my brother walks right in the hotel, service with smiles. Now the guy looks at me as if to say, "What the hell's this guy doing?" He says, "Jesus, you need a bath."

"Thank you very much. What is a bath?"

So he took me up, I don't know if it was the second or third floor, the guy says, "Here's where you sleep."

I look at the place. It's a hotel. Sheets. Beautiful bed, clean bed.

"What?" I says. "Do I sleep here?"

"Yeah, that's your room."

I took a shower. Nice. They gave me some clean clothes. Then it was time for mess. So we all went downstairs. We went down to big tables, all sitting down, they're all eating, and the guys I was sitting down with, my brother's next to me, and all the rest of the guys were shooting the bull, this and that, and I took it all in. And all of a sudden they put the food on the table.

"Jesus! Is that all we get? They give us the same old food all the time, god damn!" They're all bitching. "'What? Again we gotta eat this good food?" I mean, Food! FOOD! We ate our K rations and C rations, that was food! This was like giving you gold out of a dollar. I looked at that food, Wow! And these guys were all bitching. I says, "Why you..." I spoke up. "Why you rotten sonofabitches," just like that I told them, "You guys don't even know what the hell you're talking about." I says, "You're bitching about that food?" I says, "How would you like to eat on top of a dead body with maggots coming out of the body?" I says, "and eating C rations if you're lucky you got it." I just sat down, I says, "Thanks, fellas." I ate like a pig. "Thank you very much fellas, now get the hell outta here. I don't give a damn what you think." To me, that was food! For the first time in two years, man, chicken ... vegetables ... hot stuff. I'll tell you, I was in heaven."

- - -

Made in the USA
Monee, IL
04 November 2026

67380487R00069